Floriculture

Designing & Merchandising

Second Edition

**Delmar Publishers is proud
to support FFA activities**

Floriculture

Designing & Merchandising

Second Edition

Charles Griner

DELMAR

™

THOMSON LEARNING

Albany • Bonn • Boston • Cincinnati • Detroit • London • Madrid • Melbourne • Mexico City
New York • Pacific Grove • Paris • San Francisco • Singapore • Tokyo • Toronto • Washington

NOTICE TO THE READER

Publisher does not warrant or guarantee any of the products described herein or perform any independent analysis in connection with any of the product information contained herein. Publisher does not assume, and expressly disclaims, any obligation to obtain and include information other than that provided to it by the manufacturer.

The reader is expressly warned to consider and adopt all safety precautions that might be indicated by the activities described herein and to avoid all potential hazards. By following the instructions contained herein, the reader willingly assumes all risks in connection with such instructions.

The publisher makes no representations or warranties of any kind, including but not limited to, the warranties of fitness for particular purpose or merchantability, nor are any such representations implied with respect to the material set forth herein, and the publisher takes no responsibility with respect to such material. The publisher shall not be liable for any special, consequential, or exemplary damages resulting, in whole or in part, from the readers' use of, or reliance upon, this material.

Delmar Staff:
Business Unit Director: Susan L. Simpfenderfer
Executive Editor: Marlene McHugh Pratt
Acquisitions Editor: Zina M. Lawrence
Developmental Editor: Andrea Edwards Myers
Executive Marketing Manager: Donna J. Lewis
Channel Manager: Nigar Hale
Executive Production Manager: Wendy A Troeger
Production Assistant: Melanie T. Buck

COPYRIGHT © 2002

Delmar is a division of Thomson Learning. The Thomson Learning logo is a registered trademark use herein under license.

Printed in the United States of America
1 2 3 4 5 6 7 8 9 10 XXX 05 04 03 02 01 00

For more information, contact Delmar, 3 Columbia Circle, PO Box 15015, Albany, NY 12212-0515; or find us on the World Wide Web at http://www.delmar.com

Library of Congress Cataloging-in-Publication Data

Griner, Charles.
     Floriculture : designing & merchandising / Charles Griner.—2nd ed.
        p. cm.
     ISBN 0–7668–1560–9
     1. Flower arrangement—Juvenile literature. 2. Florists—Vocational guidance—Juvenile literature. [1. Flower arrangement. 2. Florists—Vocational guidance. 3. Vocational guidance.] I. Title.

SB449.G685 2000

745.92—dc21   00-030685

Contents

Preface

Floriculture Designing & Merchandising, 2nd Edition, was written to introduce students to the career possibilities in the floral industry and to provide basic instruction in the techniques of floral design and merchandising. It will also be useful to those students with avocational interest. Everyone can enjoy making floral arrangements for their home.

I believe that everyone can be taught to make beautiful floral arrangements and this book was written to accomplish that goal. Some of you may not be as artistically talented as others, but if you follow the step-by-step procedures outlined in this book, you will be able to make lovely arrangements that can be used in your home or given to friends. For some of you, this book will be a launching pad into further study and creativity in designing floral arrangements. Learning is a lifetime process. Should you choose a career in this area, there will always be workshops and design schools to challenge you to expand your knowledge and creative talents.

This book was written as a how-to book. It is highly illustrated to assist you in accomplishing this goal. Basic information about a wide variety of topics is presented throughout the book, but it always takes you to the how-to level of application. The designs illustrated throughout the book are simple and easy to follow. Please be aware that there are many different ways to make the same floral arrangement. Different designers have their own unique styles. Learn as many of these as you can.

The flowers used in the illustrations are the most inexpensive flowers that are readily available anywhere in the country. Feel free to substitute flowers and add your own creativity to each of the designs.

Learning objectives are presented at the beginning of each chapter to assist you in identifying what you should learn in each of the units. Self-evaluations at the end of each unit will assist you in determining if you have accomplished the objectives. Student activities are suggested to make the study interesting and to help you apply the material.

Six appendices are included to help you identify cut flowers, cut foliages, potted plants, and dried materials of importance to the floral industry. These are arranged in alphabetical order by scientific name for easy reference. The appendices also contain other pertinent information that will be useful in your study of several chapters within the book.

The information in this book can be reinforced and expanded on by your instructor and guest speakers. Put forth a dedicated effort and challenge yourself to expand your knowledge and express your creativity. Remember that flower arranging is a skill that you can learn but it will take hard work and lots of practice to become accomplished at that skill.

Acknowledgments

I am grateful to a number of individuals who have assisted me in completing this book. Foremost among these is my wife, Brenda, who spent many hours doing a variety of tasks from typing to researching information. Most of all, she and my sons, Todd and Brad, have been my encouragement.

Special thanks are due to my son Brad for drawing the original artwork for the book and to LaVon Miller and Grady Morey for their assistance with photographs.

Thanks also to Ronnie Barrett and his staff at Flowers by Barrett, Moultrie, GA, for assistance with arrangements, equipment, and pictures.

I also want to thank Denise Thompson, Tifton, GA; Doug Weeks, Napier Flower and Gift Shop; and the Stallings family at The Leaf Maker, Moultrie, GA, for their assistance.

A very special thanks to Andrea Myers of Delmar, for her endless patience and advice, and the reviewers of the manuscript for their many helpful hints and suggestions.

Brian Myers
Unity High School
Mendon, IL

Carl Vivaldi
Buck County Technical
Fairless Hills, PA

Robert Hall
Norfork High School
Norfork, AR

Louis Randall
Livingston High School
Livingston, CA

Exploring Careers in the Retail Flower Business

OBJECTIVE

To explore careers in the retail flower business.

Competencies to Be Developed

After completing this unit, you should be able to:

- begin making a decision whether you wish to explore the possibilities of a career in the retail flower business.
- identify the jobs available in the retail flower business.
- describe the duties of various jobs in the retail flower business.
- identify two types of florist.
- describe the training requirements for a job in the retail flower shop.
- prepare an employability portfolio
- identify do's and don'ts for a job interview

Terms to Know

carriage trade shop
designer
designer's assistant
employability portfolio
franchise shop
full-service shop
mass-market shop
resume
salesperson
specialty shop
stem shop
studio operation
wholesale florist
work sample

Introduction

Welcome to the exciting world of floral design. Whether you are interested in a career in the florist industry or want to learn to make floral arrangements for your own enjoyment, you are going to be challenged and rewarded by the activities included in this book. You will also develop a greater appreciation for the artistry and creativity that go into a floral design.

As you read this book, you will realize that America is becoming more conscious of flowers. Think of all the places or occasions you have seen flowers used: weddings, hospitals, parties, receptions, homes, banquets, funerals, and many others. During happy and sad times, Americans express their feelings with flowers.

The retail florist industry has been promoting the use of flowers, not just on special occasions but for regular purchase for the home. Flowers add a feeling of warmth and welcome unlike that of any other room accessory.

If you have abilities as a creative designer, you may want to consider floristry as a career. To the person who enjoys working with flowers, this career is appealing and rewarding. Not only will you be selling flowers, but you will also be selling the feelings that go with flowers. Ask a young lady how she felt when she received her first bouquet of roses. Flowers have a message—whether it be love, excitement, or sympathy—and a florist shares in the emotional response created by the flowers.

If you like working with flowers but are not a talented designer, you may want to consider a job in the florist industry other than designing.

TYPES OF FLOWER SHOPS

Different types of shops serve different groups of the population. A florist shop should tailor its services to the needs and desires of the clientele in its area. The different types of flower shops and a brief explanation of each are listed below.

Full-Service Shop

A full-service shop is the traditional retail shop offering a wide variety of services and products. Products available at a full-service shop would include fresh flowers, blooming plants, balloons, novelty giftware, and silk and dried arrangements. Such a shop also offers regular and special delivery, setups for special events, charge accounts, and wire service.

Specialty Shop

A specialty shop targets a particular need in the market by specializing in one segment of the industry. A shop may specialize in weddings, potted plants, or some other area. Specialty shops are often located adjacent to another business that services the same need in a different area. For example, a specialty wedding shop might locate next to a bridal shop.

Carriage Trade Shop

A carriage trade shop caters to an elite clientele, such as wealthy private party accounts and corporate accounts. It may offer the same services as a full-service florist, but caters to the special tastes of its clients. The prices of goods are higher, corresponding to the elite services, product lines, and designs offered. It is important for the carriage trade shop to understand the lifestyles of its clients.

Stem Shop

Stem shops are cash-and-carry operations that offer a wide variety of flowers by the stem or bunch. They do not usually offer design services or delivery. Such shops are usually located in high-traffic areas in larger cities, either inside a building or on the street.

Studio Operation

A studio operation concentrates on specialty and party work for an exclusive client base. The studio may operate out of a warehouse without a storefront, because most sales are made by appointment at the customer's home or business.

Franchise Shop

A franchise is usually purchased from a parent company and operated according to that company's rules and regulations. The prospective owner would purchase or build a shop. Sometimes the products for resale must be bought from the parent company. The price of a franchise is usually high, and the owner may have to pay a percentage of gross sales to the parent company. In spite of its higher cost, a franchise provides the business with instant recognition.

Mass-Market Shop

A mass-market shop is located in a general merchandise chain store or grocery store. These shops usually offer fresh arrangements, potted plants, balloons, and cut flowers by the stem or bunch. Mass-market shops are cash-and-carry operations and do not offer delivery service.

Wholesale Florist

The wholesale florist is an important link between the grower and the retail florist. The wholesaler purchases flowers from all around the world and supplies them to the local retail shops. In addition to cut flowers and greens, the wholesaler supplies plants, giftware, and related supplies, such as ribbon, wires, and accessories.

JOB OPPORTUNITIES IN THE RETAIL FLOWER SHOP

Retail flower shops are present in nearly all towns of any size throughout the United States. Most of these are individually owned or family-operated shops employing fewer than ten people (Figure 1-1). An employee in one of these shops may be asked to perform numerous and varied duties.

In large shops with many employees, an individual is hired to perform a specific job. These shops may be individually owned or may be part of a chain of retail shops in which each store has its own manager.

Job opportunities in the retail flower shop may be divided into three areas as listed in Figure 1-2. A description of each of these is given in the following pages.

Owner or Manager

Whether you own or manage a flower shop, your responsibilities will be similar. These include hiring, training, and supervising designers, salespersons, and delivery persons.

FIGURE 1-1

Most flower shops in the United States are small shops with fewer than ten employees.

Other responsibilities include planning work schedules, ordering supplies and flowers, coordinating sales promotion and advertising, and supervising or keeping shop records. Pricing products or training others to do so is another duty of the owner-manager (Figure 1-3).

1. Managerial and supervisory
   A. Flower shop owner
   B. Flower shop manager

2. Technical
   A. Designer's assistant
   B. Designer

3. Service workers
   A. Salesperson
   B. Office worker
   C. Delivery person

FIGURE 1-2

Jobs in retail flower shops.

FIGURE 1-3

Retail flower shop owner-manager. *Photo courtesy of M. Dzamen*

Because of the nature of their duties, owner-managers must have a good business background. In most small flower shops, the owner-manager also works as a designer and/or salesperson. Often the owner-manager may not handle routine sales but is responsible for planning and pricing weddings, receptions, and large parties.

Designer's Assistant

The designer's assistant is a person training to become a designer. Designer's assistants work with a designer and thereby are able to observe the construction of a variety of floral designs. They usually coordinate the designer's orders, keeping adequate materials on hand. The assistant may prepare containers, select and wire flowers, and construct bows for arrangements. After the designer has completed an arrangement, the assistant fills out a card or delivery tag, places it on the arrangement, and prepares the product for delivery (Figure 1-4).

As assistants gain experience, they will be allowed to make bud vases, corsages, and smaller arrangements. The difficulty of the orders given to the assistant will increase as the assistant receives further training and experience. The assistant may become a designer after 1 to 2 years.

FIGURE 1-4

Designer's assistant.

Designer

The floral designer arranges flowers and plant material in an artistic manner, and so must have an understanding of the principles of design. Designers must also have an extensive knowledge of flowers and their care, as well as the supplies used in a flower shop. They must be able to construct arrangements for the home, hospital, funeral parlor, and various special occasions.

Most shops usually designate one person as the head designer, often the manager or owner (Figure 1-5). In very large shops, someone else may be assigned this task. The head designer assigns specific work to the other designers and is usually responsible for making sales that involve special work, such as a wedding or large party.

In many shops, the head designer orders flowers and supplies. This is a difficult job because the buyer must predict in advance the quantity of materials that will be needed. If the shop runs out of flowers, it costs the shop in lost sales. If too many are ordered, they go to waste.

Designers hold an important position in the flower shop. The quality of their workmanship determines the level of

FIGURE 1-5
Floral designer.

customer satisfaction, and a happy customer means repeat business.

Salesperson

The salesperson possesses skills in the art of selling. The first encounter a customer has in a flower shop is often with a salesperson (Figure 1-6). For this reason, the salesperson's job is vital. Having the best designs in town is of little good if the salespeople are turning away customers. Customer satisfaction which leads to repeat flower buyers is necessary for a successful flower business.

In very large flower shops, salespeople may be assigned to either telephone sales or direct customer sales. However, in most shops, the salespeople perform both services. For this reason, they must have versatile and effective selling skills.

It is not realistic to think you can walk into a flower shop and instantly become a salesperson. Selling requires training and daily preparation. Salespeople must have a thorough knowledge of flowers, flower care, and floral designs. They must be able to offer customers appropriate suggestions.

In addition to having a good sales presentation, the salesperson should have a friendly attitude. Attitude communicates to the customer how you feel about yourself, the

FIGURE 1-6
Salesperson.

company you represent, and the customer. Remember, you are not only selling flowers, you are selling yourself as well.

Delivery Person

Most retail flower shops are full-service florists, offering delivery services. Delivery is a convenience to the customer and makes it possible for the customer to shop by telephone. This service distinguishes full-service florists from mass-market florists, such as those in grocery stores.

Delivery personnel should always be well groomed and well mannered (Figure 1-7). They come into contact with many people while making deliveries. The image that they present reflects upon the shop. For this reason, it is important that they make a good impression.

Duties performed by delivery personnel may include packaging orders and loading flowers into the delivery truck. They should be familiar with the delivery area so that they can make deliveries quickly and efficiently. Delivery personnel should be capable of repairing damaged arrangements so that flowers do not have to be returned to the flower shop, causing delays in delivery. In most small flower shops, delivery personnel are asked to perform other tasks in their spare time.

FIGURE 1-7
Delivery person.

TRAINING FOR A JOB IN THE RETAIL FLOWER SHOP

Most employees of retail flower shops received their training on the job. Many of these began as a salesperson or designer's assistant and worked their way up to designer, head designer, or owner-manager. Trainees in a flower shop usually participate in many of the designer schools and workshops provided by their state florist association or other groups. Even skilled designers participate in these workshops to sharpen their skills and keep abreast of the latest design techniques.

The greatest prerequisites to a career in the retail flower shop business are some artistic ability and a love for working with flowers. A good business background is also helpful. If you possess these characteristics, you can begin preparing right now in the floral design class offered in your school. Many designers today began their career in a high school horticulture class and then received additional training on the job (Figure 1-8). The national FFA organization in many

FIGURE 1-8

Many designers, such as this one, began their career in a high school horticulture class.

states offers flower shows and competitions in flower-arranging. Participating in these will help you to gain experience and further develop your flower-arranging skills. You may also want to enroll in some business classes offered in your high school.

There are a number of floral design schools throughout the nation. If you are interested in receiving further training at one of these, talk to your instructor about schools in your area. Should you decide to attend, examine the school, its curriculum, and instructors very carefully.

If you have decided to pursue a career in the florist industry, then set some goals for yourself. Do you want to work part-time while you are still in school or do you want to wait until you graduate and then seek full-time employment? Examine the employment opportunities in your area. If you live in a small town with a limited number of florists, you may have to move to a larger town where employment opportunities are greater.

LOCATING JOB OPPORTUNITIES

Once you have established your goal, you need to find a florist with a job opening. There are many ways to identify openings and employers. Your agricultural education instructor is a great place to start. Share your goals with your instructor. Many floral design teachers know the florists in the area and may know of employment opportunities.

Check the Help Wanted ads in the classified section of your local paper. But remember, only about 15 percent of all job openings are ever advertised, so do not discount a particular florist just because it has not advertised a job opening.

Go to your local state employment office. Many companies rely on them for all of their personnel resources.

You may want to check with private employment firms. They make their money by charging for their services, usually a percentage of the annual salary. Sometimes the company pays, and sometimes you pay.

Check the yellow pages of your local telephone book. Make a list of the florists in your area. Then prepare for a personal visit to the shops to inquire about job possibilities. The remainder of this chapter will help prepare you to make these inquiries.

EMPLOYABILITY PORTFOLIO

A popular trend for people seeking employment is the development of an **employability portfolio**, a collection of documents that shows you have knowledge, mastery, and job readiness in a particular occupational area. This portfolio gives your potential employer immediate proof of your qualifications.

The employability portfolio should include but not be limited to the following:

- letter of introduction
- table of contents
- letter of application
- completed job application Form
- resume
- letter of recommendation
- work samples

Letter of Introduction

The letter of introduction introduces the reviewer to the portfolio. The letter should be both personal and informative. Tell about yourself. List your personal and career goals and how you plan to reach those goals. Write about your strengths and weaknesses. Describe some of your achievements and how your work has improved.

Table of Contents

After completing all the sections of your employability portfolio, develop a table of contents to assist reviewers in locating different portfolio sections.

Letter of Application

A letter of application is written to fit a specific job with a specific employer. It states your skills and abilities as they apply to a particular job. It should be sent as a part of your portfolio or if you choose not to complete a portfolio, it should be sent with your resume. Your letter of application expresses your interest in both the job and the company.

Keep the letter brief and to the point. Sell the employer on the idea of granting you a job interview. Express your reason for contacting the company and expound on your qualifications for the job. At the closing of your letter, state a time that you will call to schedule an appointment to discuss your qualifications. See Figure 1-9 for an example of a letter of application.

Job Application Form

The job application may be a general job application (see Figure 1-10) or an actual application from industry. If you have not included one in your portfolio, you may be asked to complete a job application prior to your interview or upon arriving for the interview. Although most companies have their own version of an application, they ask basically the same questions. Come prepared to answer the questions on the application. Have a record of former employers, their addresses, and the dates you were employed. Include summer work and volunteer work if you have not been previously employed at a regular job. Also be prepared to give reasons for leaving the employment.

Fill out the job application form legibly and neatly. Print your responses on the application. Fill in every blank. If a question does not apply to you, write in N/A for "not applicable."

The job application form gives the potential employer a first impression of you. Make certain that it is a good one. A neat and complete job application form shows that you know how to follow directions and that your are thorough.

Resume

When applying for a job, you will need a resume and it should be a part of your employability portfolio. A **resume** is a list of your experiences relating to the kind of job you are seeking. There are many styles or forms of resumes. The Internet has a number of sites dedicated to resume writing and computer programs are available at many stores that sell software programs. These programs are available to help you create a resume that will attract employers' attention.

Lindsay Kilgore
133 Funston-Sale City Road
Anytown, US 31768

March 10, 2000

Mrs. Brooke Sassy
Owner
Sunny Side Florist
Anytown, US, 31768

Dear Mrs. Sassy:

I am writing to inquire about the opening your florist has for a designer's assistant. John, one of your sales persons told me about the opening today when I visited your shop and gave me your name so the I could write to you directly.

I have been enrolled in the Environmental Horticulture program at Colquitt County High School for the past two years. One of those years has been in Floral Design Class and I have decided to pursue a career in this area. I am particularly interested in working at your shop because Samantha, one of your designers talked to our class this past semester and demonstrated how to do several designs. I was most impressed with her work and her friendly attitude. I have some work experience as a sales associate in the garden center at Wal Mart.

Enclosed you will find a copy of my employability portfolio. I would like to schedule a time for an interview. Please call me at your earliest convenience to schedule an appointment. If I have not heard from you by Friday, March 14, 2000, I will give you a call.

Sincerely

Lindsay Kilgore

Enclosure

FIGURE 1-9

Sample letter of application.

APPLICATION FOR EMPLOYMENT

(PRE-EMPLOYMENT QUESTIONNAIRE)   (AN EQUAL OPPORTUNITY EMPLOYER)

PERSONAL INFORMATION

DATE _____

SOCIAL SECURITY
NUMBER _____

NAME _____

LAST	FIRST	MIDDLE

PRESENT ADDRESS _____

STREET	CITY	STATE	ZIP

PERMANENT ADDRESS _____

STREET	CITY	STATE	ZIP

PHONE NO. _____   ARE YOU 18 YEARS OR OLDER? YES ☐   NO ☐ _____

SPECIAL QUESTIONS

DO NOT ANSWER **ANY** OF THE QUESTIONS IN THIS FRAMED AREA UNLESS THE EMPLOYER HAS **CHECKED A BOX PRECEDING** A QUESTION, THEREBY INDICATING THAT THE INFORMATION IS REQUIRED FOR A BONA FIDE OCCUPATIONAL QUALIFICATION, OR DICTATED BY NATION SECURITY LAWS, OR IS NEEDED FOR OTHER LEGALLY PERMISSIBLE REASONS.

☐ Height _____ feet _____ inches

☐ Weight _____ lbs.

☐ What Foreign Language do you speak fluently? _____ Read _____ Write _____

☐ Citizen of U.S. ____ Yes ____ No

☐ Date of Birth * _____

☐ _____

*The Age Discrimination in Employment Act of 1967 prohibits discrimination on the basis of age with respect to individuals who are at least 40 but less than 70 years of age.

EMPLOYMENT DESIRED

POSITION _____

DATE YOU
CAN START _____

SALARY
DESIRED _____

ARE YOU EMPLOYED NOW? _____

IF SO MAY WE INQUIRE
OF YOUR PRESENT EMPLOYER? _____

EVER APPLIED TO THIS COMPANY BEFORE? _____   WHERE? _____   WHEN? _____

EDUCATION	NAME AND LOCATION OF SCHOOL	*NO, OF YEARS ATTENDED	*DID YOU GRADUATE?	SUBJECTS STUDIED
JR. HIGH SCHOOL/ MIDDLE SCHOOL				
HIGH SCHOOL				
COLLEGE				
TECHNICAL, BUSINESS OR CORRESPONDENCE SCHOOL				

*The Age Discrimination in Employment Act of 1967 prohibits discrimination on the basis of age with respect to individuals who are at least 40 but less than 70 years of age.

GENERAL

SUBJECTS OF SPECIAL STUDY OR RESEARCH WORK _____

MILITARY
SERVICE _____   RANK _____

PRESENT MEMBERSHIP IN
NATIONAL GUARD OR RESERVES? _____

FIGURE 1-10

Sample application for employment.

FORMER EMPLOYERS [LIST BELOW LAST FOUR EMPLOYERS, STARTING WITH LAST ONE FIRST.]

DATE MONTH AND YEAR	NAME, ADDRESS AND TELEPHONE OF EMPLOYER	SALARY	POSITION	REASON FOR LEAVING
FROM				
TO				
FROM				
TO				
FROM				
TO				
FROM				
TO				

REFERENCES: GIVE THE NAMES OF TREE PERSONS NOT RELATED TO YOU, WHOM YOU HAVE KNOWN AT LEAST ONE YEAR.

	NAME	ADDRESS TELEPHONE	BUSINESS	YEARS ACQUAINTED
1				
2				
3				

PHYSICAL RECORD:

DO YOU HAVE ANY PHYSICAL LIMITATIONS THAT PRECLUDE YOU FROM PERFORMING ANY WORK FOR WHICH YOU ARE BEING CONSIDERED?   ☐ YES   ☐ NO

PLEASE DESCRIBE

IN CASE OF
EMERGENCY NOTIFY   (   )
          NAME   ADDRESS   AREA CODE   PHONE NO.

"I CERTIFY THAT THE FACTS CONTAINED IN THIS APPLICATION ARE TRUE AND COMPLETE TO THE BEST OF MY KNOWLEDGE AND UNDERSTAND THAT, IF EMPLOYED, FALSIFIED STATEMENTS ON THIS APPLICATION SHALL BE GROUNDS FOR DISMISSAL.

I AUTHORIZE INVESTIGATION OF ALL STATEMENTS CONTAINED HEREIN AND THE REFERENCES LISTED ABOVE TO GIVE YOU ANY AND ALL INFORMATION CONCERNING MY PREVIOUS EMPLOYMENT AND ANY PERTINENT INFORMATION THEY MAY HAVE PERSONAL OR OTHERWISE, AND RELEASE ALL PARTIES FROM ALL LIABILITY FOR ANY DAMAGE THAT MAY RESULT FROM FURNISHING SAME TO YOU.

I UNDERSTAND AND AGREE THAT, IF HIRED, MY EMPLOYMENT IS FOR NO DEFINITE PERIOD AND MAY, REGARDLESS OF THE DATE OF PAYMENT OF MY WAGES AND SALARY, BE TERMINATED AT ANY TIME WITHOUT PRIOR NOTICE."

DATE   SIGNATURE

DO NOT WRITE BELOW THIS LINE

INTERVIEWED BY   DATE

HIRED: ☐ YES   ☐ NO   POSITION   DEPT.

SALARY/WAGE   DATE REPORTING TO WORK

APPROVED: 1.   2.   3.
       EMPLOYMENT MANAGER   DEPT.HEAD   GENERAL MANAGER

FIGURE 1-10

Sample application for employment (*continued*).

Most resumes are done in an outline form, but remember to keep your resume short and simple. A resume is not an autobiography. It is a sales brochure. You are selling yourself and your ability to do the job. As a general rule, include only those items that demonstrate your ability to do the work. If a resume is too long, too wordy, or too disorganized, it often gets thrown in the wastebasket. The best resumes are usually only one page in length.

There is no official format for writing a resume. You may use the following outline as a guide in completing your resume. Alter, expand, or modify it to suit your wishes.

Heading Give the employer your name and where you can be reached. At the top center of the page, type your name, address, and phone number with area code.

Job Objective State what kind of job you are seeking. Keep this to one short sentence. If you have no experience, state that you are: "Seeking an entry-level position as a designer's assistant."

Education On one line, give the date you will be graduating, the type of diploma you will be receiving, the school's name, and the city-state address. Under the name of your school, list a few of the courses you have taken that would help qualify you for the job.

Skills and Abilities This area may or may not be a part of your resume. If you do not have any work experience, then listing any skills or abilities that would qualify you for the job will be helpful. List any workshops, trade shows, seminars or self-study courses that you have taken.

Work Experience If you have work experience—full-time, part-time, volunteer, or charitable—put it in your resume. List your most recent job first. On one line, list the date you left that job or write "Present" if you are still employed, your job title, the name of the company, and the company's address. Directly under that line, briefly describe the duties you performed while you were in this job.

Extracurricular Activities Extracurricular activities show leadership capabilities, team spirit, interest, and experiences that you have had. If you do not have work experience, these can be particularly important.

References References are people who can give information about you, your attitudes, and your abilities. Try to select people who know you from different activities. Be sure to ask the individuals for their permission to use them as references. For each reference, you will need the name, job title, organization, address, and telephone number.

The notation "References available upon request" is often included on the resume instead of listing individual references. However, you always need to have the list available in case it is requested.

If you are using a word processor on your computer to type your resume, choose a font that is easy to read. Use only standard abbreviations and spell out any acronyms. For example instead of using FCCLA, spell it out: Future Career and Community Leaders of America. Examine the sample resume in Figure 1-11, to see how a completed resume should look.

Letter of Recommendation

Your employability portfolio should contain one or more letters of recommendation. At least one letter should be from a credible source that has knowledge of your specific skills. Other letters of recommendation may be from other sources to represent your character traits and other personal or academic accomplishments.

Work Sample

Your **work sample** should showcase your best work. Take pictures of your work and add a caption describing the work sample, steps taken to complete the work sample, skills demonstrated by your work sample, and what you learned or how you could improve the work sample.

FOLLOW-UP CALL

The follow-up call is most important. Shop owners-managers receive many inquiries about the availability of jobs and they are probably not going to call you. They figure

Lindsay Kilgore
2222 South Main Street
Anytown, US 00000
021 941 5894

OBJECTIVE: Seeking an entry level position in Floral Industry

EDUCATION: Colquitt County High School, Anytown US

- Floral Design Program
- Expected Date of Graduation, June 2000
- Relevant Courses: Floral Design
                          Floriculture

SKILLS: Floral Design
- Know procedures for conditioning cut flowers and greens.
- Construct boutonnieres and corsages.
- Construct basic floral arrangements.

Floriculture
- Can identify 35 different plants sold in floral shops.
- Can advise customers on plant care.

WORK: 2000, Wal Mart Garden Center, Anytown, US

Position: Sales Associate (Summer hire)
Assisted customers and cared for plants in the Garden Center.

1999, Colquitt County Regional Hospital
Delivered flowers to patients and ran errands (Volunteer Work).

EXTRA-CURRICULAR ACTIVITIES: Vice-President, Colquitt County FFA Association
First place, state FFA floral design contest
Future Career and Community Leaders of America (FCCLA)

REFERENCES: Available on request

FIGURE 1-11
Sample resume.

that if you really want to work for them, you will do more than send them a copy of your employability portfolio or resume.

Practice what you are going to say before you call. Be friendly and speak clearly. Do not be put off if the manager says there is no opening at the present time. Ask if he or she can meet with you and discuss your qualifications for future job openings. Sometimes, shop owners will create a position if the right person comes along.

THE INTERVIEW

The interview is an opportunity for you to meet with a representative of the florist to which you have applied. It is a chance for you to sell yourself. If you are going to do well on the interview, you must show the interviewer that you are capable of doing the job or learning it quickly, and that you are reliable and do not need constant supervision. Interviewers also look for workers with a good attitude who can work well with other people in the shop.

Interviews at most flower shops are usually very informal, but they can also be scary. Prepare ahead of time so that you do well on the interview. Make a list of potential questions the interviewer might ask. Here are some suggestions:

- Tell me about yourself.
- Tell me about your education.
- Tell me about your previous jobs.
- What are your goals?
- Why do you want this job?
- What are your strengths?
- What are your weaknesses?
- How would you (or someone else) describe your personality?
- With so many applicants, why should I hire you?
- What do you plan to be doing in five years? Ten years?
- Do you plan to continue your education?
- Why did you leave your previous job?
- Why would you like to work here?
- What did you like and dislike about your last job?
- Can you work well under pressure?
- How long will you stay with us?

- Do you have any questions?

Seek the advice of others who have been on interviews to help you compile your list of questions. Then practice your answers. Ask a friend to play the role of manager. Practice until you are relaxed and confidant with your answers. Now you are ready for your interview. The following do's and don'ts will help you make a good impression.

Do:

- Be clean and neatly dressed. Appropriate dress is the same as that worn on the job.
- Learn all that you can about the florist and the job for which you are applying.
- Be prepared to talk about yourself—your accomplishments, skill, and abilities.
- Arrive a few minutes early.
- Be alert and enthusiastic.
- Be friendly.
- Maintain good eye contact.
- Speak clearly and in a pleasant voice.
- Answer questions as completely and truthfully as possible.
- Maintain good posture.
- Ask questions about the job.

Don't

- Be late.
- Chew gum, smoke, or eat candy.
- Slouch or slump when sitting.
- Bring a friend with you to the interview.
- Be dishonest when answering questions.
- Ask about salary and benefits. Wait for the manager to bring up the issue of wages.

Interviewing for a job with a floral shop can be different than interviewing for most other jobs. You may be asked to perform certain skills such as making a corsage or an arrangement. This can be scary but it also gives you the opportunity to showcase what you can do.

When the interview is about over, offer a closing statement. Summarize your strengths and ask for the job. Many people interview and never ask for the job.

Also ask when you might expect to hear from the interviewer. Regardless of the outcome, smile and thank the manager for talking with you. Give a firm handshake. Ask when you might call and inquire about the decision.

AFTER THE INTERVIEW

It is good manners after the interview to send a typed follow-up letter. This letter could make you stand out from the competition and it keeps your name alive in the manager's mind. It also tells the manager that you really want the job. Write the letter as soon as you get home from the interview. If you have not heard from the manager in a few days, call to inquire about the position. If he or she has not made a decision, this lets the manager know that you are still interested.

SUMMARY

Most retail flower shops are small businesses so jobs within the shop are not usually limited to one area. An employee may be asked to perform a variety of tasks.

A career in the retail flower business is an excellent choice for those individuals who enjoy working around flowers and plants. Work in a retail florist is highly seasonal, with flowers in especially great demand on holidays. Employees are expected to work long hours during these times. Because most weddings take place on weekends, employees may be asked to work longer hours during these events also.

Even though flower shop employees are often required to work long hours, most florists love their work and receive great personal satisfaction from their job.

Student Activities

1. Visit a flower shop and question the employees about their jobs.
2. Invite the owner-manager or designer from a local flower shop to talk to the class about careers in the retail florist industry.

3. Select one job opportunity and make a list of the duties performed on that job and the requirements for employment in that area.

4. Write about a situation where you used flowers or saw flowers being used. How did they make you feel?

5. Prepare an employability portfolio for a specific job.

6. Make a list of possible questions that might be asked on an interview.

7. Role play a job interview situation letting one student play the role of the interviewer and one the role of the job applicant.

Self-Evaluation

A. True or False

_____ 1. Most retail flower shops employ ten or more individuals.

_____ 2. People employed in large flower shops usually perform one specific job.

_____ 3. Three areas of employment in the retail flower shop are managerial and supervisory, technical, and service-oriented.

_____ 4. Wedding arrangements are designed by the designer's assistant.

_____ 5. A skilled designer must possess a great deal of artistic ability.

_____ 6. Selling in the retail flower shop requires both training and daily preparation.

_____ 7. As long as a salesperson is trained in selling, a knowledge about flowers is not necessary to be successful in the retail flower shop.

_____ 8. Flower arrangements damaged during delivery should be returned to the flower shop for repair.

_____ 9. Employees in small retail flower shops must be able to perform a variety of duties.

_____ 10. Employees in a retail flower shop can expect to work regular hours.

_____ 11. A flower shop that specializes solely in wedding flowers would be an example of a studio operation.

_____ 12. The wholesale florist sells flowers and supplies to the retail florist.

_____ 13. Eighty-five percent of all job openings are advertised in the classified section of the newspaper.

_____ 14. Private employment agencies receive a percent of the job applicant's annual salary as their fee.

_____ 15. The best resumes are no more than two pages in length.

_____ 16. Volunteer work should not be included on the resume.

_____ 17. Interviews are usually informal and do not require any preparation.

_____ 18. After the interview, the job applicant should not annoy the manager with telephone calls but rather wait to be called.

B. Short Answer Questions

1. Why is it important that the delivery person for a flower shop be well-groomed and well-mannered?

2. What is the difference between a full-service florist and a mass-market florist?

3. How can the National FFA help you train for a job in the retail flower shop?

4. Why is a friendly service attitude an important attribute of a salesperson?

5. How can designers gain more experience in floral design and learn the latest techniques?

6. List five items that should be included on the resume.

7. List five do's and five don'ts of interviewing.

The History of Floral Design

OBJECTIVE

To relate how designs from historical periods influence contemporary designs.

Competencies to Be Developed

After completing this unit, you should be able to:

- identify the origins of many of the floral design that are popular today.
- identify the characteristics of mass, line-mass, and line designs.

Introduction

Flower arrangement is considered an art. How do we arrange flowers so that they become a work of art? In Unit 3 we learn that there are design principles that guide us in creating arrangements that are works of art. What are these principles and where did they originate? They are fundamental truths proved by master artists over the centuries. In order to understand the principles of design and how they originated, we need to look at the history of the art of flower arranging.

Terms to Know

American federal
  period
baroque period
Byzantine period
chaplets
Colonial
  Williamsburg
  period
cornucopia
Dutch-Flemish period
Early American period
Egyptian period
Empire period
English-Georgian
  period
faience
French baroque period
French period
French rococo period
gothic period
Greek period
line designs

(continued)

line-mass designs
Louis XVI period
mass designs
Middle Ages
nosegay
occidental design
oriental design
Renaissance period
Roman period
Victorian period

A knowledge of the floral arts of earlier cultures is important to better understand the flower arrangement styles used today. Also designers are sometimes required to create flower arrangements that depict a specific period or style of design. For example, a customer with a Victorian home would want arrangements that complement the Victorian period in history.

A study of the history of floral design reveals that two different concepts of floral design developed independently of each other. Our sources of inspiration have been the **occidental style**, which evolved in Egypt and was further developed by the Europeans, and the **oriental style**, which began in China and was later developed by the Japanese.

CLASSICAL PERIOD

The remains of many ancient cultures provide us with ample proof that people have always appreciated the beauties of flowers. They show that flowers were a source of inspiration for decorating the home or were presented as offerings.

Egyptian Period (2800–28 B.C.)

Ample evidence exists to show that the ancient Egyptians decorated with cut flowers placed in vases. The usual Egyptian container was a basin or a wide-mouth bowl that tapered to a narrow base. Such bowls of gold, silver, pottery, and **faience** (a ware made of finely ground silicate) were fitted with devices for holding flowers and fruit.

Floral arrangements of Egypt were simplistic, repetitious, and highly stylized. Flowers were set in regimented rows around the edge of the vase, an ample 2 inches above the rim. These blossoms were flanked by leaves or buds on slightly lower stems. There was no bunching or overlapping of material.

Faience bowls for flowers had holes around the rim through which flowers were inserted. Blossoms in tall spout vases came straight from the opening with no stems visible.

The primary colors of red, yellow, and blue as well as other vibrant color combinations were used predominately in floral designs. Because the lotus, or water lily, was the flower of the goddess Isis, and therefore considered sacred, it was used often. Bowls of fruit and flowers were also used but always in orderly sequence of color and shape.

The Egyptians liked to wear wreaths of flowers as well as flower collars and made chaplets for their hair. They liked to carry bouquets of flowers made of lotuses with buds and blossoms of other flowers threaded into them.

Clarity and simplicity were characteristic of Egyptian floral designs. The Egyptians did not care for confusion or complexity. Two artistic functions prevailed in these designs: repetition and alternation—one flower around the rim of a vase and alternation of color, blue followed by green, then by blue again.

Greek Period (600–46 B.C.)

The ancient Greeks were so dedicated to beauty that their art heritage has lived through the ages and influences today's art. However the Greeks did not arrange their flowers in vases or bouquets. Figure 2-1 has been reproduced from a

FIGURE 2-1

The Greeks placed branches of foliage into vases but did not include flowers.

very rare Grecian example of decorated terracotta showing plant material in vases. Yet flowers are lacking here as in all Greek examples. The leafy branches are probably of olive for this is part of a bridal scene and the olive was associated with weddings.

Flowers were commonly scattered on the ground during festivals and used to make garlands worn around the neck and wreaths, or **chaplets** worn on the head. Because the wreath was the symbol of allegiance and dedication, it was awarded in honor to athletes, poets, civic leaders, soldiers, and heroes.

Funeral graves were decorated with garlands of flowers as were banquet tables. Wreaths were so much a part of the Greek way of life that books were written to describe the appropriate flowers, forms, and etiquette for wearing them.

Greek designs expressed grace and simplicity. Color was not important. Instead, the flowers, fragrance, and symbolism associated with each flower were foremost importance. Flowers were often symbolic of a god or hero.

The **cornucopia** or horn of plenty was first introduced by the Greeks (Figure 2-2). Originally it was placed in an up-

FIGURE 2-2
The Cornucopia.

right position whereas today it is laid on its side with contents overflowing and spilling out. As the symbol for abundance, we often associate it with our Thanksgiving celebrations.

Roman Period (28 B.C.–A.D. 325)

The Romans contributed little that was new in the use of flowers but continued the customs of the Greeks. Wealth and power, however, led the Romans to greater luxury in the use of flowers and they were used in abundance at religious rites and banquets. At banquets roses were strewn on the floor to a depth of 2 feet, and flowers "rained" from the ceiling. The fragrance of so many flowers was said to be suffocating. So customary was the use of roses at the evening meal that it was called "the hour of the rose."

One custom that appears in Roman art is the use of scarves for carrying flowers. Flowers were carried on a scarf and offered at an altar as a part of Roman religious ceremonies. The use of wreaths and garlands was continued from the Greeks. However, the Roman wreaths and garlands were heavy and elaborate. Wreaths, like high crowns, came to a point over the forehead. Garlands were even more elaborate, being wide in the center and tapering toward the ends.

There is evidence that the Romans actually arranged flowers in baskets. These baskets were high at the back and flattened in front. The flowers were placed low between feathery branches so that the flowers were clearly visible. This required adequate spacing of the branches. The flowers used in these arrangements were highly fragrant and bright in color.

Byzantine Period (A.D. 320–600)

In the fourth century, Byzantium was chosen by the first Christian emperor, Constantine, as the Eastern capital of the Roman Empire. Soon afterward the Western Roman Empire was overrun by barbaric tribes and entered that period of turmoil that we call the Dark Ages. Byzantium was able to protect itself from foreign conquest until 1453 when it was captured by the Ottoman empire of the Turks.

Greek and Roman flower usage styles were continued, but the garland was constructed differently. The background was of foliage into which tiny flowers were set in arching lines to give a twisted effect.

Byzantine flower compositions were distinguished by height and symmetry. Containers were filled with foliage to resemble symmetrical, conical trees. These were decorated at regular intervals with clusters of flowers or fruit (Figure 2-3).

FIGURE 2-3

This conical tree design as inspired by the conical designs of the Byzantine period.

EUROPEAN PERIODS OF FLORAL DESIGN

Several important periods of floral history have influenced European floral art. As you study these periods, you will begin to see floral styles that have influenced the styles popular today.

Middle Ages (A.D. 476–1400)

The centuries between the fall of Rome in 476 and the dawn of the Renaissance in the fifteenth century are called the **Middle Ages**. These were years of unrest and confusion in Europe. Order and security were slow to emerge.

We know little of the uses of flowers in Europe from the seventh century to the thirteenth. Monks grew herbs for medicine and fruits and vegetables for eating. Flower gardening as such did not exist. We know, however, that monks were familiar with many wildflowers for glimpses of them appear in the manuscripts over which they worked.

During the later part of the Middle Ages, known as the gothic period, flowers began to take a more important role in daily life. Borders of manuscripts and altar pictures and their frames blossomed with painted plants and flowers.

Renaissance (A.D. 1400–1600)

Renaissance period saw a rebirth of many interests, particularly in the arts. The Renaissance began in Italy but quickly spread to all of Europe. The Renaissance style was greatly influenced by the Byzantine, Greek, and Roman periods. Flowers in vases are often shown in paintings from this period as great emphasis was placed on flower symbolism.

Characteristic floral arrangements of the Renaissance were flowers arranged in vases so that only the blossoms were visible. Stems were covered creating a massed, symmetrically stiff arrangement. Even though the flowers were compactly arranged, each flower stood out because of the variety of bright colors and forms of flowers that were used.

The Renaissance was given to pageants and festivals and artists were commissioned to design floral pieces for them. Fruits, blossoms and leaves were woven into garlands to decorate walls and vaulted ceilings. Petals were piled into baskets to strew on floors and streets or to float down from balconies into rooms below.

Many traditional floral designs created today are styled from the Renaissance arrangements such as the Christmas wreath of fruit, cones, and flowers.

Baroque Period (A.D. 1600–1775)

The baroque style, like that of the Renaissance, originated in Italy and spread to the rest of Europe. In the works of Michelangelo and Tintoretto we see examples of this new style emerging. By 1650 baroque arrangements could be seen in paintings and tapestries of the period.

Early in the **baroque period**, arrangements were typically massed and overflowing. They were often created as symmetrical, oval-shaped designs. Later in the period, asymmetrical curves in the shape of an S or a crescent became popular. The S curve (Figure 2-4) was created by an English

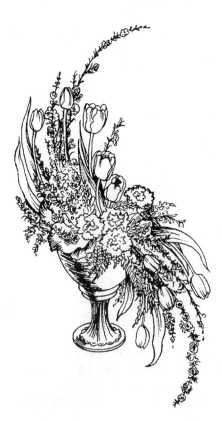

FIGURE 2-4

The S curve design became popular during the Baroque period.

painter named William Hogarth. The Hogarthian curve is still quite popular in modern floral designs.

Dutch-Flemish Period (A.D. 1600–1750)

We also gain insight into the baroque style of arrangement through flower paintings by the Dutch and Flemish artists. Traditional baroque styles were refined as they created floral designs for their paintings. These arrangements were not as loose and open as in the contemporary baroque style, but they were better proportioned and more compact. A major distinguishing characteristic of this period is the great variety of flowers within one bouquet.

It is important to understand that many of these early artists never actually arranged the flower bouquets or painted their pictures using an actual arrangement. This explains why short-stemmed flowers are often positioned high and flowers from all seasons are placed side by side.

French Period (A.D. 1600–1814)

The French styles for decorating changed often during this period of history. Four basic periods are discussed here.

The first period is called **French Baroque**. This period occurred during the seventeenth century during the reign of King Louis XIV. The French Baroque style was directly influenced by traditional Baroque art. However, certain features made it purely French in origin. The court society during the reign of Louis XIV had become idle and effeminate with extravagant tastes for luxury. Feminine appeal became an important characteristic of the floral designs of this period in France. The emphasis in flower arrangements was on refinement and elegance as compared to the flamboyance of the Dutch-Flemish period.

The second artistic period in France is know as the **French Rococo**. This style began in France but quickly spread throughout Europe and the European colonies. This change in style occurred during the reign of King Louis XV. The informal designs of the French Baroque gave way to the more formal, feminine designs that are characteristic of the French Rococo period. These floral arrangements were predominantly asymmetrical and curvilinear in form with the crescent (C curve) used more often than the Hogarthian curve (S curve).

Flowers used in Rococo designs were delicate and airy. The predominant colors were subtle rather than contrasting.

The third artistic period in France is referred to as **Louis XVI** (late eighteenth century). This period showed a continued movement to femininity in design styles. This was brought about by Queen Marie Antoinette who favored delicate, cool colors highlighted with gold.

The fourth artistic period in France was called the **Empire Period** (1804–1814). Following the French Revolution in 1789, a new artistic movement evolved across Europe known as the classical revival period or neoclassical period.

Nowhere else in the Western world were neoclassical styles as they were during the rule of Napoleon Bonaparte in France. Under the guidance of two of his architects, the Empire design style was created. These were masculine designs characterized by militaristic themes. Femininity was dropped from French design. Empire arrangements were massive in size and weight. They were more compact than those of earlier French periods with simple lines in a triangular shape and strong color contrasts. A typical Empire design would be arranged in a heavy urn containing an abundance of large, richly colored flowers.

English Georgian Period (A.D. 1714–1760)

The **English-Georgian** period in England was named after the three English rulers King Georges I, II, and III. These kings ruled England during the baroque period. Most English-Georgian arrangements were formal and symmetrical, often tightly arranged with great varieties of flowers. During this period, floral designs were greatly influenced by the Chinese arts because of active trading between Europe and the Orient. The Chinese style was incorporated into Georgian arrangements by the creation of symmetrical forms, usually triangular-shaped floral designs.

During the later years of the Georgian period, floral designs moved away from formality and symmetry. The fragrance of flowers became important because it was believed that their perfume would rid the air of diseases. Because of this belief, the English created the **nosegay**, a small hand-held bouquet to carry the sweet scents. Nosegays also helped mask the smells of body odors in a society where bathing often was not believed to be healthy.

These small handheld bouquets are often called **tussie-mussies**, sometimes spelled tuzzy-muzzy. The word tuzzy refers to the old English word for a knot of flowers. These bouquets were first used solely for fragrance but soon became a fashion trend. Women of the Georgian period wore flowers in their hair, around their necks, and on their gowns.

Victorian Period (A.D. 1820–1901)

Flowers were considered fashionable during the **Victorian period**. This period was named after Queen Victoria of England. However, floral designs during this era were generally poorly proportioned. Large masses of flowers were placed tightly into a container to create a compact arrangement. A typical flower arrangement would have an asymmetrical balance and a massed, tightly compact effect. No definite style of arrangement was prevalent. Designs were often a blending of the art of previous periods. So many different colors and flowers were used that the arrangement appeared unplanned.

The nosegay, introduced during the English-Georgian period, was very popular. These bouquets were used as air fresheners and the flowers conveyed special sentiment as well.

Toward the end of the Victorian period, attempts were made to establish rules for floral arranging. The art of flower arranging was taught by skilled designers. Hence, flower arranging became a professional art.

AMERICAN PERIODS OF FLORAL DESIGN

The early settlers in America brought with them a European heritage of floral design. As the settlements became established and trade began to take place, the arts began to emerge.

Early American Period (A.D. 1620–1720)

The early setters brought the styles of the Renaissance with them to America, but life was hard in the new colonies and they had little time to devote to art and flower arranging. The colonists were gardeners, but their attention was focused on providing plants and herbs for food and medicine.

As the early settlements became established, the colonists placed wildflowers, grains, and grasses into everyday jars, simple pottery, pewter and copper kettles and pans.

Colonial Williamsburg Period (1714–1780)

By the time Williamsburg became the capital of the Virginia colonies, active trading was taking place with England, Europe, and Asia. The artistic styles from these areas were adapted into the New World art. The typical floral arrangement of this period was a massed, rounded, or fan-shaped bouquet that was casual and open in style. The arrangements were constructed so that the flowers were lightly arranged at the top, while flowers with greater visual weight were placed above the rim of the container.

American Federal Period (A.D. 1780–1820)

The **American federal period** was equivalent to the English-Georgian period in England. This period was greatly influenced by the neoclassic and Empire designs that evolved in Europe at that time. The colonies had just received their independence from England and the American people wanted to break away from the traditions of England.

The shape of the arrangement from this period was often pyramidal or fan-shaped, influenced by the French design style. The floral designs were little different from those found in the neoclassic movement and gradually gave way to the ornate and stuffy design of the Victorian period.

Nineteenth-Century American Flower Arrangements

The artistic styles of American flower arrangements changed little during the early part of the twentieth century. The Victorian era was coming to a close by the end of World War I. Flower arrangement styles were copies of preceding periods

FIGURE 2-5

European design was generally a large, round or oval mass of flowers.

or blends of several design styles. The corsage became popular in the 1920s to be worn for special occasions. This custom has survived to the present time.

Major changes were brought about in American floral art at the end of World War II. This occurred because of a renewed interest in Japanese culture. While continental Europe continued the tradition of the loosely arranged **mass designs** (Figure 2-5), American flower arrangements incorporated the **line-mass** style (Figure 2-6). The line-mass designs combined both oriental and European ideas. American floral design used more materials than the oriental design, but far fewer than the European. American floral designs were often built around a linear pattern, further showing the oriental influence.

ORIENTAL FLOWER ARRANGING

The Oriental style of flower arranging actually began in India where Buddhist priests scattered branches and stems on altars or placed them in pottery urns as decorations. The practice was quickly picked up and modified by the Chinese priests during the first century A.D. They arranged the flowers in massive bronze ceremonial vessels, and because they

FIGURE 2-6

A line mass design combines Oriental and European ideas.

felt it improper to place flowers carelessly on the altar, they created symbolic arrangements. Chinese arrangements were usually large and symmetrical, with only one or two types of foliage and flowers placed around a central branch or main axis. Bright colors contrasting with the color of the urn were favored. The flowers having the lightest colors were used at the outer portions of the design, while darker ones were kept nearest the base.

Around the sixth century A.D., the Japanese adopted many aspects of the Chinese culture, including that of floral arrangement. A Japanese Buddhist priest, named Ikenabo, refined that art and ritual, and his instruction was sought by other Buddhist priests. He is credited with having begun the first school of floral art in Japan, which bears his name— Ikenabo. This school still exists. The name was later changed to ikebana, which means "giving life to flowers." Many other schools of Japanese floral design have evolved from this original one, but the basic principles can be traced back to the teachings of the Ikenabo school.

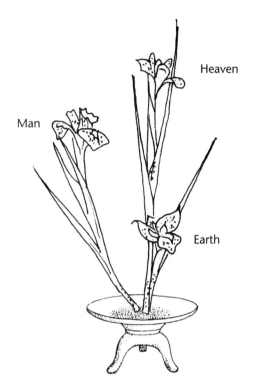

Heaven

Man

Earth

FIGURE 2-7

Oriental design is characterized by minimum use of plant materials and each placement has meaning.

The Japanese designs are characterized by minimum use of plant material and the careful placement of branches and flowers. Each placement has meaning as does the angle of placement (Figure 2-7). This type of design came to be known as **line arrangement**. If you have an interest in this style of design, ask your instructors if they have any books on Ikebana or check with your local library.

Student Activities

1. Select pictures of period arrangements, from interior and home furnishings magazines. Make a poster-board display of the arrangements, identifying the artistic period.

2. Divide the class into small groups. Assign each group an artistic period and let them prepare an oral presentation on that period with emphasis on floral arrangements of the period.

Self-Evaluation

A. Fill in the blank space with the answer that best completes each statement.

1. The oriental and _____ concepts of floral design developed independently of each other.

2. Three terms that best describe Egyptian flower arrangements are simplistic, repetitious, and _____.

3. The _____ was considered sacred to the Egyptians.

4. The Greeks and Romans used garlands and _____ extensively.

5. The rebirth of interest in the arts following the Middle Ages is known as the _____.

6. The English painter, William Hogarth, created the _____ design.

7. Four artistic periods in France include French Baroque, _____ Louis XVI, and the _____ period.

8. The _____ design was popular during the French Rococo period.

9. During the English-Georgian period, people believed that the perfume of flowers would rid the air of _____.

10. Small handheld bouquets during the English-Georgian period were sometimes called _____.

B. Match the following floral design with the period in which it originated.

_____ 1. Chaplets

_____ 2. Conical trees

_____ 3. Tussie-mussies

_____ 4. Hogarth S design

_____ 5. Heavy masculine

_____ 6. Cornucopia

_____ 7. Christmas wreath

A. Byzantine period

B. Renaissance period

C. Baroque period

D. English-Georgian period

E. Greek period

F. Empire period

C. Short Answer Questions

1. Explain how the Oriental and European floral design styles have influenced the types of flower shop arrangements sold today.

2. Why is it important for floral designers to have a knowledge of the history of floral design?

Principles of Design

OBJECTIVE

To utilize principles and elements of design to critique floral designs.

Competencies to Be Developed

After completing this unit, you should be able to:

- list and define the principles and elements of design.
- identify the basic designs used in flower arranging.
- identify the basic color schemes used in floral design.
- use a color wheel to determine combinations for various color schemes.
- critique an arrangement using a rating scale based on the principles of design studied in this unit.

Introduction

If you were given a dozen roses or carnations in a gift box, you would probably place them casually in a vase. Grouped together this way, they are beautiful because individual flowers are beautiful. But this casual placement of flowers is not an arrangement. These same flowers placed in a vase in a planned pattern take on a greater beauty: The flower arrangement then becomes a work of art.

We arrange flowers so that they become a work of art by using certain guidelines called *principles of **design***. These are basic laws, fundamental truths, or methods of operation that have been tested and proved by master artists for many centuries. Good flower arrangements are judged by these principles. These principles are tools that will guide you in planning and evaluating your arrangements.

Although these interrelated principles have been developed over many years and are used as guidelines in all types of designs, artists individualize their application. Such variations depend on many factors. For examples, the choice to employ one principle may affect the way several other principles are expressed in the arrangement.

DESIGN PRINCIPLES

The design principles presented in this unit are balance, proportion and scale, focal point, emphasis, rhythm, harmony, and unity.

Balance

Balance refers to the stability of an arrangement. When all of the design elements are composed so that the arrangement appears secure and stable, then balance has been achieved.

Balance must be both visual and actual. Visual balance refers to the way an arrangement appears to the eye. It is achieved by the proper use of color and the placement of plant materials according to size.

Visual weight refers to how heavy an object appears in the arrangement. Flowers of dark colors appear to be heavier than flowers of lighter colors even if they are the same size. Likewise, flowers or objects of coarse texture appear heavier than similarly sized objects of smooth texture. Because darker colored and coarsely textured flowers appear to be heavy, they must be placed near the base of the arrangement. Lighter, smoother ones are placed near the edges. This helps give the arrangement visual balance.

Actual balance or mechanical balance is achieved by the proper placement of flowers so that there is an equal amount of weight on both sides of a central axis in the arrangement. If mechanical balance is not achieved, then the arrangement may topple.

split-complementary
 harmony
texture
tint
tone
transition
triadic harmony
unity
value

Materials List

floral design
 magazines
floral arrangements
a selection of flowers
 differing in color
 and texture

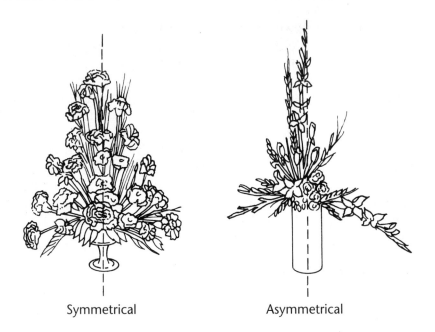

FIGURE 3-1

Balance.

Symmetrical   Asymmetrical

Two kinds of balance, symmetrical and asymmetrical, are commonly used (Figure 3-1). Symmetrical or formal balance is characterized by equal visual weight on each side of an imaginary, central, vertical line. This equal visual weight does not have to be expressed in identical materials, but the materials are usually very similar. More traditional styles of arrangements are based on this type of balance.

Arrangements that are symmetrically balanced give a feeling of dignity and formality. They are poised rather than moving, passive rather than active. Symmetrical arrangements should be displayed against a symmetrical background and accessories displayed in a symmetrical way. Arrangements placed before the altar in a church or on the head table at a banquet hall are usually symmetrically balanced.

Asymmetrical or informal arrangements have equal visual weight on both sides of a central axis, but each side is different in plant materials and the manner of arrangement. Japanese styles of arranging are based on asymmetrical balance and have greatly influenced our contemporary style of arranging. Asymmetrical balance is active rather than passive and suggests movement to the eye. This type of balance is more informal.

Proportion and Scale

Most arrangements are designed for a particular location such as a dining table. The size of the dining table and the colors in the room will determine the flowers used, the size and shape of the arrangement, and the container chosen. This relationship is known as proportion and scale.

Proportion is the interrelationship of all parts of an arrangement—flowers, foliage, accessories, and container. Floral designers generally agree on the following proportion principles: the plant materials should be 1-1/2 times as high as the height of a tall container or 1-1/2 the width of a low container (Figure 3-2).

These are minimum dimensions. The maximum dimensions depend on the background space and the weight of the materials to be used. If the materials to be used are light and airy, the arrangement may be two to three times or more the height of the container. Also, due to their actual and visual weight, containers made of sturdy materials and dark colors can hold much larger arrangements.

Select plant materials that are consistent in size and character with the container and each other. If a delicate, crystal container is chosen, use flowers in keeping with the fragile nature of the crystal. Large, heavy mums would not be appropriate. A better choice would be roses or carnations. Also, select flowers which are alike in size. Tiny flowers, such as baby's breath, would be a poor choice with large dahlias.

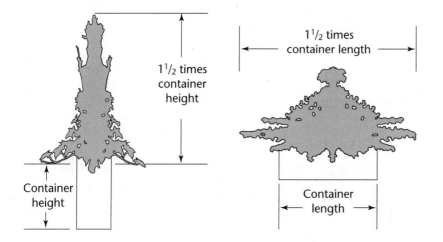

1¹/₂ times container height

Container height

1¹/₂ times container length

Container length

FIGURE 3-2
Proportion.

Scale refers to the relationship between an arrangement and the area where it is to be displayed. A heavy oak dining table in a large room would need an arrangement of considerable weight and mass. Fragile flowers would look out of place in such a setting. A large arrangement placed before the altar of a church illustrates good scale.

Focal Point

The **focal point** of a design is the area that attracts and holds the interest of the viewer. This spot or area dominates the design. The focal point, also called the "center of interest," is located near the place where the stems or main lines appear to meet (Figure 3-3). The radiating lines direct the eye to the

FIGURE 3-3
Focal point.

center of interest. Balance requires that the center of interest be near the base of the design.

The strength of the focal point is dictated by the style of the arrangement. A round arrangement placed on the center of a table does not really have a focal point since the arrangement must be equally attractive from any position around the table. Many modern arrangements require a strong focal point, while the focal point of line arrangements should be moderately strong because the lines of the arrangement should dominate.

There should be only a single focal point in a design. Arrangements with more than one create a restless movement of the eye, and unity within the arrangement is destroyed. The following are suggestions for creating a focal point.

1. Bring the main lines of the design to the point.
2. Place the largest flower there.
3. Concentrate the plant material in that area.
4. Place the darkest color or the brightest color there.
5. Place strongly contrasting colors or textures at this point.
6. Place an unusually shaped flower there.

Emphasis

Emphasis and focal point are closely related. Emphasis focuses the attention on one feature and keeps everything else secondary. The focal point is one way of creating emphasis in an arrangement, but other factors are also involved. Texture, color and kind of flower, and movement combine to achieve it as well. Good use of emphasis or dominance establishes order within the arrangement.

It is easy to develop a design using different kinds of flowers, foliage, and colors. When the arrangement is finished, each of these compete for dominance and the arrangement becomes uninteresting. Use a predominance of one color, texture, line, or kind of flower and complement this with small amounts of other colors and flowers. By maintaining a dominance of one type of material, the finished design is more pleasing.

Rhythm

Rhythm is the movement of the eye through a design toward or away from the center of interest. It is the flow of lines, textures, and colors that evokes a sense of motion. The viewer's eye should move back and forth smoothly between the focal point and the outer edges of the arrangement. Rhythm can be created by repetition, radiation, progression, and transition.

Repetition is the simplest way to develop rhythm in a flower arrangement. Repetition is accomplished by repeating the leading color, the strongest line, the dominant form, or the dominant texture in an arrangement. To achieve repetition, select a flower of the desired color and repeat the use of this flower throughout the arrangement. The container is also a part of the finished design, and the colors and textures of the container may be repeated also. For example, crystal agrees in texture with delicate flowers, while heavy pottery is suitable for coarse flowers.

Radiation is an attempt to make all stems appear to come from one central axis (Figure 3-4). The point of origin should be the focal point of the design. This helps create an emphasis and a strong sense of unity in the arrangement.

Progression in an arrangement involves a gradual change by increasing or decreasing one or more qualities, including the size, color, or texture of the material used, or the space between flowers. Through progression, we develop movement in a certain direction.

FIGURE 3-4
Radiation.

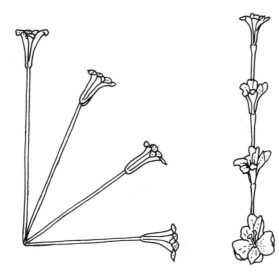

FIGURE 3-5

Facing flowers to achieve a progression in size.

Progression in size is accomplished by using flowers of increasing size. Place buds and small flowers at the edge of the arrangement, large flowers at the bottom and center of the arrangement, and medium-sized flowers between them. If all the flowers are the same size, face them in different directions to alter their "visual weight." **Facing** flowers increases or decreases the eye appeal or visual weight (Figure 3-5).

Do not confuse facing flowers with selecting the "face" of a flower. Many flowers have one side that is more appealing than the other (Figure 3-6). When placing a flower in an arrangement or corsage, place the most attractive side of the flower, the face, toward the front of the design.

Progression in space can be achieved by increasing the space between flowers at the edges of an arrangement and decreasing it at the center. (Figure 3-7).

Progression in color can be accomplished in much the same manner. Use flowers of light value at the top and edges, dark values at the center of interest, and intermediate values between the two.

To achieve progression in texture, proceed from materials with fine texture through medium texture to coarse texture. Finely textured materials are placed at the edges of the arrangement, while coarse ones are near the center of interest.

Transition also involves making a gradual change, allowing the designer to harmonize unlike things. Transition

FIGURE 3-6

Selecting the face
of a flower.

refers to the blending of colors, line patterns, and textures. Transition and progression are closely related, and progression is often used to create a transition in the design.

Avoid sectioning your design. Sectioning occurs when you use one color or texture in one area and a different color or texture in another. Blend colors, textures, and shapes together to unify the design.

Transition should also exist between the container and the arrangement. Allow some of the plant material to overlap the rim of the container. By doing this, the eye moves easily from the container to the arrangement.

Harmony

Harmony refers to a blending of all components of the design. If a design is harmonious, there should be a pleasing relationship within the design in color, texture, shape, size, and line so that a clear, central idea of theme exists. Disregard of a design principle leads to a lack of harmony.

Unity

Unity is achieved when all the parts of the design suggest a oneness in idea or impression. Unity is achieved, in part, by repeating the colors throughout the design. The establishment of a focal point and a dominant flower in the arrangement also helps to create unity. Do not layer flowers, colors, or textures in horizontal rows within a design. Layering destroys unity. Good transition in an arrangement helps to achieve unity by leading the eye smoothly from one part of the arrangement to another.

DESIGN ELEMENTS

The principles of design are like a recipe with the elements of design as the ingredients. These elements are the visual qualities of a composition—line, form, texture, and color.

FIGURE 3-7
Space rhythm.

Line

Line provides a visual path for the eye to follow, thus creating motion in the design. This line is the framework that holds the entire arrangement together. The designer creates line by using linear materials, such as stems, branches, or line flowers. Line flowers are long, slender spikes of blossoms with florets blooming along the stem. Line may also be developed by the placement of round flowers in sequence, creating a feeling of direction.

Line in a design sets an emotional tone. It can give the viewer feelings of swift motion, repose, reverence, and gentleness. Vertical lines imply strength, dignity, or feelings of formality, while a curved line adds gentleness. The curved line adds a feminine dimension. Horizontal lines suggest informality and make people feel restful. For this reason, the horizontal line is often used for table arrangements.

To maintain movement, the lines of a design must never be broken. If they are, a restless feeling results. Lines should also appear to originate from one point. Line movement in the most common designs can be seen in Figure 3-8.

Form

Form can be defined as the shape or silhouette of an arrangement. The design might have a circular or triangular form, or might be composed of a number of curved lines. As discussed earlier in this chapter, these design forms developed around the basic types of arrangements originating in two different parts of the world.

Geometric shapes, circles, and triangles, developed from the European type of mass arrangements. The linear shapes evolved from the Oriental type of designs. A combination of these two types of designs gives us the line mass forms common to America. Figure 3-8 shows examples of the basic design forms and their geometric shapes or line movement. As a beginning designer, become skilled in the basic ones. Later, express your own creativity by exploring variations.

Texture

Texture refers to the surface appearance of flowers, foliage, container, and accessories, such as ribbons and balloons.

Oval

Inverted-T

Vertical

Asymmetrical Triangle

Horizontal

Hogarth Curve

Right Angle

Round

Crescent

Diagonal

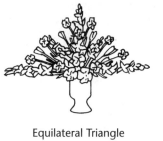

Equilateral Triangle

Fan

FIGURE 3-8

Design forms and their geometric shapes or linear movement.

FINE TEXTURE	MEDIUM TEXTURE	COARSE TEXTURE
Acacia	Pompon	Yarrow
Alstroemeria lily	Carnation	Zinnia
Snapdragons	Gerbera daisy	Dahlia
Rose	Stock	Protea
Lily-of-the-valley	Liatris	Chrysanthemum
Gardenia	Salal	Pine
Orchids	Camellia leaves	Holly
Plumosus	Spruce	

FIGURE 3-9

Examples of flowers and foliages of differing textures.

The textures of each of these can be fine or coarse, smooth or rough, shiny or dull, satiny or velvety. A rose has a fine, smooth, shiny surface while a zinnia has a coarse, rough appearance.

Generally, texture is designated as fine, medium, or coarse. Examples of flowers and foliage having each of these textures are listed in Figure 3-9.

Textures can also be used to create an emotional response from the viewer. Fine textures and smooth, shiny surfaces give the appearance of elegance or formality. Coarse textures and rough or dull surfaces create a sense of informality and would be appropriate for casual occasions. Similarly, rough textures seem strong and masculine while fine ones are elegant and feminine.

Flowers and foliages of similar textures are usually used together but contrast in textures can be used depending upon the result the designer is trying to achieve. Contrasting textures usually call attention to themselves and can be used for that purpose. Contrast also adds interest to an arrangement.

Color

Color is probably the single most important element of floral design. An arrangement can be designed beautifully, but if the colors are not pleasing, the arrangement will not appeal to a customer. Fortunately, it is only necessary to understand a few basic principles to make effective use of color.

The Color Wheel The **color wheel** is a tool that may help you understand the use of color. If you hold a prism up to the sunlight, colors of the rainbow will appear. These colors make up the colors of clear light that can be seen by the eye. This is known as the visible light spectrum and includes red, orange, yellow, green, blue, and violet. When you look at a rainbow, these colors always appear in this order. A simple color wheel can be constructed by placing them in a circle.

The traditional color wheel is made up of twelve colors (see color insert). The six colors that comprise the visible light spectrum are called **primary** and **secondary** colors. The other six **intermediate**, or **tertiary**, colors are created when primary colors are mixed in equal amounts with an adjacent secondary color. These colors are also called hues. All originate from three primary colors—red, yellow and blue, which cannot be made by mixing any other combinations of colors.

When one mixes two primary colors, secondary colors are created. Red combined with yellow produces secondary orange. Yellow plus blue yields secondary green. Blue added to red produces secondary violet (see color insert).

Other color characteristics important to the designer are chroma and value. **Chroma** is a measure of brightness or dullness and is determined by the amount of pigment in a flower. Pigments are minerals or other chemicals that reflect light in specific ways so that we see color. They absorb certain colors and reflect others which the eye detects. A yellow marigold contains yellow pigment, so it absorbs all other color and reflects yellow.

Value Value is how light or dark a color is. Colors are lightened by the addition of white, muted by gray, and darkened by adding black. Add white to a color and create a **tint**. Add gray for a **tone**, and black to create a **shade** (see color insert). By altering the value, a designer controls the emotional impact of a design.

There are rules for applying color theory. For example, lighter tints and tones should be used in greater amounts than darker ones. The rule of thumb suggests that in an arrangement containing three colors, 65 percent of the flowers should be of the lightest value, 25 percent the mid-range, and 10 percent the darkest. Use the lightest colors predominantly on the edges, and the middle values next nearest the center. Blend the darkest value into the focal area. This does not suggest that you create layers of color in an arrangement. The colors should be interspersed in such a manner that a sense of unity is achieved.

Emotional Responses to Color   Colors create an emotional response that is different for each person. The study of how individuals react to color is called the psychology of color. This reaction can usually be traced to one's educational background, personality, and geographical location. For example, what colors do you associate with Christmas? Your immediate response was probably red and green. Red and green are colors that we have been taught through association to relate to Christmas. This may not be true in other geographic areas.

Colors may appear to advance (toward) or recede (from the viewer). This is illustrated in the colors we use to paint a room. A large room may be painted in warm, advancing colors to make it appear smaller and cozier. A small room might be painted in cool, receding colors such as blue or green so the room appears larger. This response to colors is important when selecting colors for arrangements. If an arrangement is to be viewed from a distance as might be the case in a large building, such as a church, select flowers in warm, advancing colors such as yellow. Flowers of a cool hue tend to fade out and cannot be seen from a long distance.

Colors can also create moods, such as excitement or relaxation. In general, outgoing people are attracted to warm colors: red, orange, and yellow. Private people are usually attracted to cool colors: green, blue, and violet. Additionally, each of the colors in flowers or foliage conveys a symbolic message to the viewer which can, in turn, create a theme or emotional response to an arrangement. Some of these are listed below.

yellow—happy, cheerful, bright, symbolic of friendship and the spring season.
red—stimulating, exciting, warm, joyful, expresses love.
blue—quiet, cool and retiring, dignity and formality.
orange—warmth, autumn color, pumpkins and Halloween.
green—restful, symbolic of living things and St. Patrick's Day.
violet—restful, denotes royalty and elegance.

black—somberness, death, Halloween, dramatic effect.
white—purity and innocence, weddings, snow.

Color Harmonies Color harmonies are combinations of
colors that are pleasing to the eye. The designer creates these
by grouping different hues or combining tints, tones, and
shades of a single hue.

A **monochromatic** color scheme consists of a single hue
and its variations in tint, tone, and shade (Figure 3-10 and
see color insert). One value should dominate for interest. A
monochromatic design will become monotonous unless the
materials have interesting form and texture. Color values
must also be selected with care. Different flowers of the same
value would be harmonious but lack interest. Add a flower of
darker or lighter value and the arrangement develops greater
appeal.

An **analogous harmony** is achieved by using three or
more hues which occur next to each other on the color wheel
(Figure 3-11 and see color insert). One of these should be a
primary color, which should predominate over the others.

A **complementary** color harmony is a combination of
any two colors opposite each other on the color wheel (Figure 3-12 and see color insert). This combination produces a
strong contrast of cool and warm colors. Cool colors, grouped
on one side of the color wheel, include violet, blue, green,
and their intermediate colors. Warm colors are grouped on

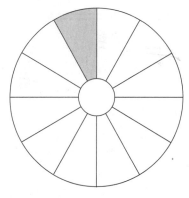

FIGURE 3-10

A monochromatic color
harmony.

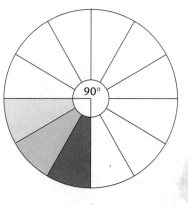

FIGURE 3-11

An analogous color harmony.

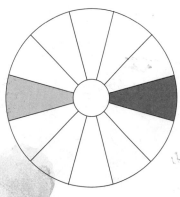

FIGURE 3-12

A complementary color
harmony.

the other side and include yellow, orange, and red, and their intermediate hues. Use complementary colors to dominate the arrangement and create an emphasis. Use tints and shades of the complementary color to accent the design.

A **split-complementary** color scheme uses any color with the two colors on each side of its complement (Figure 3-13 and see color insert). The color contrast is not as great for this harmony as in the complementary color scheme. Select a single color for emphasis.

A **triadic** color harmony is found by using any three colors that are equally spaced on the color wheel (Figure 3-14 and see color insert). Allow one color to provide emphasis.

A polychromatic color harmony uses three or more unrelated colors (Figure 3-15 and see color insert). When using this color scheme, select tints and shades that are pleasing together. Again, allow one color to dominate.

GENERAL GUIDELINES

The principles of design are interrelated, and a designer cannot alter one principle without affecting another. For example, an arrangement that lacks unity will probably not be harmonious. Flower arranging is an art, and while every designer has a unique style of arranging flowers, each must adhere to the principles of design. Apply the principles to examine your designs and those of others.

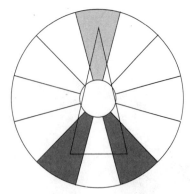

FIGURE 3-13

A split-complementary color harmony.

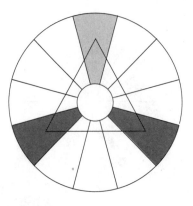

FIGURE 3-14

A triadic color harmony.

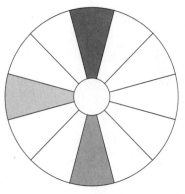

FIGURE 3-15

A polychromatic color harmony.

1. A flower arrangement should be about 1-1/2 to 2 times the height or length of the container. The horizontal arrangement is an exception to this rule.

2. Use no more than two or three kinds of flowers.

3. Use no more than three colors.

4. Use an uneven number of flowers when the total to be used is less than twelve.

5. Large flowers and dark colors should be used toward the bottom and center of the arrangement.

6. Small flowers should be used toward the top and edge of the arrangement.

7. If the flowers are of equal size, there should be about three light flowers for every dark flower.

8. Establish the height of a design first; the length second; and the width third.

9. Flowers should not crowd or touch each other. Leave space around the flowers.

10. Place some flowers deep into the arrangement to create depth.

11. The focal point should be prominent.

12. Each type of flower should form the design shape as it is placed in the arrangement.

13. Break the rim of the container with a flower or a leaf to tie them together.

FIGURE 3-16

General guidelines for flower arranging.

Designers have developed a list of guidelines to help them adhere to the principles outlined in this unit (Figure 3-16). These will help avoid making some of the most common mistakes in flower arranging. Please note that these are only guidelines, and there will be exceptions to them.

Student Activities

1. Look for pictures of arrangements from old catalogs or magazines, and classify them according to the following criteria.

   a. Type of design—line, line mass, mass.

   b. The pattern of design—horizontal, circle, right angle, etc.

c. The color harmony used in the design.

d. Type of balance—symmetrical or asymmetrical.

e. Does the arrangement have a focal point?

f. What flower in the arrangement has the greatest emphasis?

2. Ask your instructor to display an arrangement to the class. Use the Flower Arrangement Rating Scale (Appendix G) to evaluate it. Compare your evaluation with other members of the class.

3. Visit a local florist and apply design criteria to various arrangements.

4. Invite a local florist to visit the class as a guest speaker.

5. Plan an FFA Flower Show at the end of the class.

6. Volunteer the class to make arrangements for banquets held at the school.

7. Make a color wheel using the flower illustrations from old seed catalogs.

8. Select pictures of arrangements from old magazines. Discuss how the flower colors create a specific mood and how the arrangement might be used.

Self-Evaluation

A. Select the answer that best completes each of the following statements.

1. The distribution of "visual weight" on either side of a vertical axis is known as

a. harmony.

b. balance.

c. unity.

d. focal point.

2. A crescent design is an example of

a. symmetrical balance.

b. nonsymmetrical balance.

c. asymmetrical balance.

d. semicircular balance.

3. The focal point of a design is located
   a. at the outside edge.
   b. at the top.
   c. high in the middle portion.
   d. in the lower center.

4. A small arrangement on a large buffet would not be in proper
   a. scale.
   b. proportion.
   c. balance.
   d. rhythm.

5. The visual path the eye follows to create motion in a design is known as
   a. texture.
   b. line.
   c. pattern.
   d. proportion.

6. A zinnia is an example of a flower having a _____ texture.
   a. fine
   b. medium
   c. shiny
   d. coarse

7. The primary colors are
   a. red, yellow, and blue.
   b. orange, green, and violet.
   c. red, green, and blue.
   d. blue, orange, and violet.

8. Of the following colors, _____ shows the highest value.
   a. burgundy
   b. maroon
   c. hot pink
   d. red

9. The blending of all components of a design is known as
   a. progression.
   b. transition.
   c. unity.
   d. harmony.

B. Match the color harmonies below with the example that illustrates each color harmony.

_____ 1. monochromatic
_____ 2. analogous
_____ 3. complementary
_____ 4. split-complementary
_____ 5. triadic

a. apricot, coral, tangerine
b. orange and blue
c. yellow, blue violet, red violet
d. orange, green, violet
e. red, red orange, orange

C. Match each of the following colors with the appropriate emotional response or color association.

_____ 1. red
_____ 2. green
_____ 3. blue
_____ 4. yellow
_____ 5. orange
_____ 6. black
_____ 7. violet
_____ 8. white

a. purity
b. happiness and cheerfulness
c. dignity and formality
d. love
e. autumn, Halloween
f. royalty and elegance
g. somberness
h. St. Patrick's Day

D. Short Answer Questions

1. Explain the difference between actual and visual weight as it relates to balance.
2. Describe how a designer can create a focal point in an arrangement.
3. Explain how the lines of an arrangement create an emotional response from the viewer.

Selection of Cut Flowers and Greens

OBJECTIVE

To identify the most commonly used cut flowers and foliage.

Competencies to Be Developed

After completing this unit, you should be able to:

- classify flowers according to four groups based on form and shape.
- identify the most commonly used cut flowers.
- identify the most commonly used foliages.
- identify the availability of flowers and foliages.
- describe the method of pricing and packaging of the most commonly used flowers and foliages.
- identify the colors in which flowers are available.
- identify the keeping quality of flowers.

Terms to Know

filler flowers
form flowers
line flowers
mass flowers

Materials List

examples from the four groups of flowers
examples of as many of the listed flowers as possible

Introduction

Ordering flowers is a major task for the manager of a retail florist. The manager must have considerable knowledge about the flowers: their availability, colors, keeping quality, and usual price. If the manager orders too many flowers, then his losses will be great, cutting store profits. If not enough flowers are purchased, then the designer can not readily fill orders.

FIGURE 4-1

The wholesale florist purchases flowers from growers and sells them to retail florists. *Photo courtesy of M. Dzamen*

Most retail florists order their flowers from wholesale florists, who have a wide variety available and make regular deliveries to retail stores (Figure 4-1). Many retail florists also buy directly from growers. These florists must know where growers are located and how the flowers may be shipped. Growers of flowers and foliage are located all over the world so this can be a big task.

FLOWER AND FOLIAGE CLASSIFICATION

Each piece of plant material used in an arrangement has distinguishing characteristics. As a designer, you will need to know these characteristics and how the flowers best fit together. The success of any arrangement depends on the designer's ability to select the correct flowers and foliage.

The form of the plant material usually will determine the design or pattern of the arrangement. Because of this, consideration must be given to the characteristics when selecting plant material for any given arrangement. Plant material is classified into four groups based on form and shape.

Line Flowers

Line flowers are long, slender spikes of blossoms with florets blooming along the stem (Figure 4-2). Bare twigs and other

FIGURE 4-2
Line flowers.

similar material are also classified as line flowers. These flowers are used to establish the skeleton or outline of an arrangement. They determine the height and width relationship of the arrangement. Examples of line flowers include gladiolus, snapdragons, and stock. Eucalyptus is an example of a linear foliage.

Mass Flowers

Mass flowers are single stem flowers with large, rounded heads (Figure 4-3). This group of flowers is used within the framework of the linear flowers toward the focal point. If line flowers are not available, then use buds and smaller mass flowers to create the framework of the design. Vary the heights and depths of mass flowers so that each flower stands out and the arrangement has depth. Examples of mass flowers include roses, carnations, and daisies.

Filler Flowers

Filler flowers are used to fill in the gaps between mass flowers and give depth to the design (Figure 4-4). They follow the patterns set by the line and mass flowers and add emphasis to the main blossoms.

FIGURE 4-3
Mass flowers.

FIGURE 4-4
Filler flowers.

Filler flowers may be of two types, bunchy or feathery. Bunchy flowers, such as pompon chrysanthemums, have many small mass-type heads. Feathery fillers such as gypsophila (baby's breath) give a delicate, feminine appearance to an arrangement.

Form Flowers

Form flowers have unusual, distinctive shapes (Figure 4-5). Because they add emphasis to an arrangement, they are ideal to create a focal point in a design. Space form flowers in a design so that they maintain their individuality. Never bunch form flowers together. Examples of form flowers are orchids and calla lilies.

Each of the four classes of flowers has an important place in the creation of an arrangement. Flowers from any one group or combination of the groups can be used. As a designer, you must be able to identify many flowers from each group and know the characteristics of each flower. Some of the most commonly used flowers and cut foliages are listed in Appendix A and B respectively. Other important information is also included with each.

FIGURE 4-5
Form flowers.

Student Activities

1. Take a class field trip to a wholesale florist. Ask the manager to share where he orders flowers and how they reach his shop.

2. Invite the manager of a wholesale or retail florist to talk to the class about the varied flowers and foliage used.

3. Prepare a display of flowers and foliage for identification purposes. Continually change the display during the class as flowers and foliages become available. Label the flowers by name and classification.

4. Practice preparing an order to a wholesale florist for the following number of flowers:

   15 snapdragons   8 stems of statice

   36 carnations   12 stems of baby's breath

   16 stems of pompons

5. Hold a plant identification contest. Award small prizes or privileges to the winner.

6. Select an arrangement from a magazine. List each of the flowers in the arrangement and classify them as either line, mass, filler, or form flowers.

Self-Evaluation

A. Select the best answer from the choices offered to complete the statement or answer the question.

1. When ordering flowers from a wholesale florist, one should consider
   a. availability.
   b. keeping quality.
   c. price.
   d. all of the above.
   e. a and b.

2. Plant material is classified into four groups based on form and shape. These are

a. line flowers, foliage, mass flowers, and tropicals.

b. mass flowers, filler flowers, foliages, and form flowers.

c. line flowers, mass forms, filler flowers, and form flowers.

d. none of the above.

3. All of the following are line flowers except

a. gladiolus.

b. daisies.

c. stock.

d. snapdragons.

4. All of the following are mass flowers except

a. orchids.

b. carnations.

c. roses.

d. daisies.

5. Form flowers

a. have unusual, distinctive shapes.

b. should never be bunched together.

c. add emphasis to an arrangement.

d. all of the above.

e. a and c.

B. Match the following flowers with their method of packaging.

_____ a.	alstromeria	1.	individually
_____ b.	carnation	2.	10/bunch
_____ c.	cattleya orchid	3.	25/bunch
_____ d.	gardenia	4.	amount varies
_____ e.	gladiolus		
_____ f.	China aster		
_____ g.	roses		
_____ h.	baby's breath		
_____ i.	pompons		
_____ j.	statice		

C. Short Answer Questions

1. What is the role of the wholesale florist?
2. List the four classifications of flowers and examples of each.
3. Explain how line, mass, filler, and form flowers are used in an arrangement.

Conditioning and Storing Cut Flowers and Greens

OBJECTIVE

To receive a shipment of flowers and treat them in a manner that extends the keeping quality of the flowers.

Competencies to Be Developed

After completing this unit, you should be able to:

- identify the causes of premature flower deterioration.
- identify the steps in handling a shipment of flowers.
- demonstrate proper stem treatment.
- describe the benefits of floral preservatives.
- identify storing requirements of flowers and greens.
- identify how flowers should be handled in the home.

Introduction

Having flowers that last for a long period of time is important to the florist. Long-lasting flowers please customers, making them want to return to the florist when they need flowers in the future. If flowers die too quickly, clients may seek a new florist. As a florist or as a consumer of flowers, you need to be informed about the proper treatment of flowers to make them last longer.

Terms to Know

antitranspirants
bactericide
botrytis
conditioning flowers
ethylene gas
hydration
pH
photosynthesis
respiration
stomata
succulent
total dissolved solids
transpiration
xylem

Materials List

*assorted flowers as
    they are received
    from wholesaler*
flower containers
floral preservative
knife

FIGURE 5-1

The Chain of Life symbol represents all segments of the floral industry doing their part to extend the life of fresh flowers and plants.

When a customer purchases flowers from a local flower shop, the flowers have gone through a chain of distribution. This chain may include growers, shippers, wholesalers, and the retail florist. Each of the stages in the chain can affect the quality of the flowers. For this reason, the Chain of Life Program was developed by the Society of American Florists (Figure 5-1). This program helps growers, wholesalers, and retailers lengthen the life of flowers by providing information on proper care and handling throughout the marketing chain.

Whether a business is a member of the Chain of Life Program or not, a thorough knowledge of flower care and handling benefits everyone. It results in better, longer-lasting flowers. It means extra enjoyment for the customer and that translates into profit for the retail florist.

COMMON REASONS FOR EARLY FLOWER DETERIORATION

Refrigeration and water are two commonly known requirements for making flowers last. Many florists put them in water and set them in the cooler. That is good, but not good enough. To better understand how to make flowers last longer, it will be helpful to know what causes flowers to deteriorate early. Figure 5-2 lists many of the conditions that lead to early flower deterioration and remedies for each.

Five of the most common causes of early flower deterioration are presented here with a brief explanation of each.

Low Water Absorption. Most flower stems are at least partially blocked when they arrive at the retail florist. This blockage can have many causes. The stems may have been cut with dull equipment or with shears that pinch the **xylem**, or water-conducting tubes, in the stem. Sometimes bacteria or minerals in the water clog the stem. Air can also enter the stems at the time of cutting and cause a partial blockage, which can become so severe that flowers wilt in their container.

Loss of Water. Flowers lose water through their leaves by a process called **transpiration**. Transpiration is the process by which gases, including water vapor, move from an area of greater concentration to an area of lesser concentration. Since a turgid flower contains a greater concentration of water vapor than the air surrounding it, the water naturally moves out into the air, through **stomata**, the tiny openings on the bottom of the leaves. Flowers wilt because moisture is lost

CONDITION	PROBLEM CAUSED	REMEDY
Alkalinity	microorganisms grow faster, acid water moves through plant faster	use special preservative for alkaline water to lower pH to 3.5–4.
Bacteria & fungi	clogs vascular system; decays flowers; produces ethylene	use clean buckets; lower pH; keep cool; keep leaves out of water; keep cooler clean; replace scratched buckets; recut stems
Hard water	keeps preservative from lowering pH	use special preservative for hard water; have water tested; do not use water softener
High TDS (Total Dissolved Solids)	keeps preservative from lowering pH	have water tested and matched to special preservative; may need to use deionizing procedure if TDS is over 200 ppm
Air	keeps water from moving up stem	cut stems under water; use warm water (100–105° F); soak foam properly; do not "pull back" on stems when inserting
Low nutrition	flowers starve	use preservative; replenish preservative water every 2 days
High temp	increases microorganism growth, flower metabolism	store near 32° F (check recommendations); keep out of sunlight
Drying out	wilting, burning of flowers	keep away from drafts, heat; maintain preservative water level; maintain humidity level
Low humidity	flowers dry out; wilt	maintain 90% humidity in cooler; use a "floral" cooler; cover greens & susceptible flowers; use Crowning Glory
Ethylene	causes rapid aging	use STS; separate ethylene sensitive flowers; lower temperature; clean cooler & buckets; do not store with fruits & vegetables; remove old or damaged flowers; protect from auto exhaust & faulty heaters
Improper conditioning	causes wilting; short life; drying out	condition by putting in 105° F preservative water at room temp overnight then put in the cooler several hours before use

Prepared by Dr. Frank Flanders, Agricultural Education, The University of Georgia.

FIGURE 5-2

Preventing early flower deterioration.

through transpiration quicker than it is taken in through the stems. This natural loss of water occurs more rapidly at higher temperatures and low humidity.

Loss of Food. Even though flowers have been cut, they are still living and must have a source of food. Food stored in the leaves and stem may be used up if the flowers are stored in the dark for long periods of time. Light is important in the production of food in the leaves of the plant through a process called **photosynthesis**. This process continues after the flowers are cut if given the proper light and a source of sugar.

Disease. Diseases in the cooler can cause damage or loss of flowers. Of particular importance is **botrytis**, a fungus which causes brown spots on petals. When preparing flowers for the cooler, do not allow the blooms to get wet. If they do, allow them to dry before placing them in the cooler.

Ethylene Gas. **Ethylene gas** is a naturally occurring gas in flowers that hastens maturity. Because it hastens maturity, ethylene causes rapid deterioration of cut flowers.

There are many sources of the gas. The flowers themselves, especially when diseased or injured, give off ethylene. Fruit, especially apples, also produces ethylene gas. Rotting foliage below the water line is another source, as well as exhaust fumes from automobiles.

The symptoms of ethylene gas are premature death, flower and petal drop, yellowing of foliage, loss of foliage, and upward cupping of petals, known as "sleepiness" in carnations. By avoiding the sources of ethylene, the retail florist can avoid the damage it causes. Fruit should never be stored with cut flowers. Old, diseased, or injured flowers should be removed from the cooler, and the cooler should be kept clean. Cut flowers should never be subjected to exhaust fumes from delivery trucks.

WATER QUALITY AND FLOWER DETERIORATION

Flowers contain a network of tiny vessels called capillaries, which are like tiny drinking straws. These capillaries carry water and nutrients up the stem to the leaves and flowers. This process is called hydration. **Hydration** is what keeps cut flowers fresh.

Many things can prevent hydration from taking place. Several of these have been discussed in this chapter. Another contributing cause of poor hydration is water quality.

All water is not the same. Various waters may look alike yet be very different. A simple water test will reveal the character and quality of the water.

Two important characteristics of water are pH and total dissolved solids (TDS). The **pH** is a measure of how acidic or basic a water is, on a scale of 0 to 14. Seven on the scale is neutral. All readings below seven are acidic, while readings above seven are basic, or alkaline. Recent data suggests that a low pH between 3.2 and 4.5 maximizes hydration. **Floral preservative**, materials added to water to prolong flower life, generally lower the pH levels, enabling better hydration to take place.

Another characteristic of water is the amount of **total dissolved solids** (TDS), a measure of the dissolved salt and minerals in the water. Not all minerals in the water are bad. Just as people prefer drinking water with some minerals in it, flowers generally do better in water with controlled levels of selected minerals, rather than pure water or very salty water.

Companies that produce floral preservative often formulate a number of them. One of these companies, Floralife, Inc., produces three formulations for use with various pH and TDS levels. Floralife Original Formula was designed for most average water qualities and yields optimum effect in about 70 percent of all water.

Special Blend Hard Water was created for extremely hard or alkaline waters, such as well water or water from areas high in limestone. Special Blend Pure Water was formulated for use with purified or naturally pure waters.

Floralife will perform a complete water analysis for the retail florist and make a recommendation for the right fresh flower food.

CONDITIONING FLOWERS

Having discovered the causes of premature flower deterioration, we are now ready to learn the techniques to counteract these causes (Figure 5-3). By counteracting these causes, we can prolong the life of cut flowers. These techniques of treating flowers to extend their life are known as **conditioning**.

WHAT TO DO WHEN FLOWERS ARRIVE FROM THE WHOLESALER

1. Wash buckets with bleach or special floral cleaner.
2. Mix preservative at recommended rate in 100–105° water.
3. Inspect for damage, overheating, decay, etc.
4. Groom, remove damaged flowers or foliage.
5. Wash shears or knife and recut stems under preservative water. Cut stems at a slant.
6. Remove all foliage that will be under water.
7. Place in preservative water.
8. Remove sleeves except on roses.
9. Allow to sit out 1–2 hours at room temperature (if dry packed, allow them to sit out 8–10 hours)
10. Place in cooler for several hours before use in arrangements.

Prepared by Dr. Frank Flanders, Agricultural Education, The University of Georgia.

FIGURE 5-3

Conditioning flowers.

Flowers and foliage delivered by a wholesale florist are shipped in large cardboard boxes and wrapped in paper for protection (Figure 5-4). Usually a small bag of ice is also included in the box to help prevent heat buildup. However, the flowers are shipped dry and may have been in the delivery truck for many hours. The following discussion will examine the steps in conditioning a shipment of flowers.

1. *Unpack the flowers.* As soon as flowers arrive, unpack them and loosen the paper or plastic sleeves in which they have been wrapped. As flowers mature they expand. If the sleeves are not loosened, many of the flowers will be crushed as the blooms open. Do not loosen the sleeves on roses because the customers desire their flowers in the bud stage.

Check flowers for signs of disease, damage, or wilting. Remember, diseased and damaged flowers give off an increased amount of ethylene gas which accelerates flower deterioration. If the number of diseased and damaged flowers is minor, remove them from the bunch before storage. If the damage is excessive, it should be reported to the supplier.

FIGURE 5-4

A flower shop employee unpacks a shipment of flowers.

2. *Recut the stems.* The stems of all flowers should be recut before being placed in storage, because of the blockage that occurs in the end of the stem, as discussed earlier in this chapter.

Stems should be cut with a knife rather than shears. Even sharp shears can pinch the xylem tubes causing a partial blockage. Make the second cut one to two inches above the stem base. Cut the stems on a slant. This will not help the stem absorb more water. Rather, an angle cut prevents the stems from sealing to the bottom of the container as they would with a straight cut.

Cutting the stems under warm water is beneficial (Figure 5-5). When cut in the air, the stem may suck up a small amount of air, forming a bubble at the base which can create a blockage. Use warm water because it contains less air than cold does. After the stems have been cut, they can be moved to storage containers. A clinging drop of water on the end of the stem will protect the new cut. Stems that are difficult to cut under water by hand may be cut using an underwater cutter (Figure 5-6). This device cuts fresh flowers

FIGURE 5-5

Recut stems with a sharp knife under warm water.

underwater by the bunch. It has a blade that sits in a tub and automatically cuts all stems on an angle.

The stems of some flowers may require special treatment. The flowers of woody stems should be cut like other flowers, but instead of cutting an inch off the bottom, move up the stem to the part that is more **succulent**. Succulent stems are softer and juicier and therefore better conductors of water.

Stems of plants such as poinsettias have a milky sap that bleeds after being cut. This sap reseals the cut surface and prevents the stem from absorbing water. To prevent this, burn the end of the stem over a small flame until the stem is blackened (Figure 5-7). Another method is to place the end of the freshly cut stems in a container of boiling water for 10 to 30 seconds. Take care that the heat does not damage the foliage or flowers. This procedure causes the sap to clot and allows the stem to absorb water.

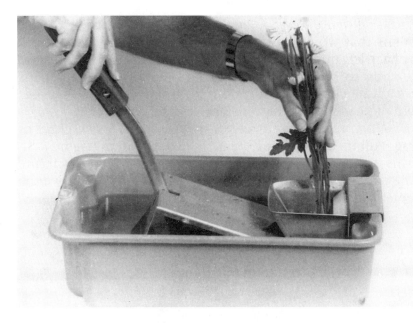

FIGURE 5-6
An underwater stem cutter.

FIGURE 5-7
Burn the end of the stems containing a milky sap over a small flame until the stem is blackened.

3. *Remove lower foliage.* Remove all foliage from the stems that would be underwater in the storage containers. Foliage left under the water will decay, fouling the water and leading to bacterial growth. This results in clogged stems which block water absorption. Rotting foliage also releases ethylene gas which hastens flower deterioration.

When removing foliage or thorns, use a heavy glove or rag. Pull the glove or rag quickly down the stem (Figure 5-8). This does a faster, neater job than a knife, which can injure the bark of the flower. If roses are being treated, remove the outside petals of the rose also.

FIGURE 5-8

Use a heavy glove or rag to remove foliage or thorns.

4.　*Clean containers and cooler.* Once flowers have been cut and cleaned, they are ready for storage. Use containers that have been scrubbed with a hot detergent solution, disinfected with bleach, and rinsed thoroughly. A 10 percent solution of bleach can be used to disinfect the containers.

Floralife, Inc., makes a product, Floralife Formula D.C.D., that can be used to clean and disinfect in one step. D.C.D. stands for "disinfects, cleans, and deodorizes." This material kills many bacteria, fungi, and algae which produce ethylene gas and also block flower stems. In addition, the detergents clean and deodorize container surfaces as they are disinfected.

Regardless of the product used, sanitation is the key. Diligently scrub buckets, shelves, floors, and sinks, and remove all sources of ethylene gas. These vital steps will prolong the life of your fresh flowers.

Nonmetallic containers should be used. Metal containers can decrease preservative effectiveness. To prevent flower injury, select a container that is short enough that the flowers will not come in contact with the sides of it.

5.　*Use a preservative.* Place a warm preservative solution in the container prior to adding flowers. The temperature of the solution should be between 100 and 110° F. This temperature should feel very warm to your hands but not burn. Warm water should be used because it contains less air and will travel up the stem more easily.

Preservatives added to flower water have been found to be very effective in extending the life of cut flowers. They do this in three ways.

First, they provide a food source. As long as the flower is alive, it continues to convert stored food reserves into energy. This process is called **respiration**. When the stored food in the cut flower is used, the flower dies. The preservatives contain sugar which the flower uses to manufacture food to replace that lost through respiration.

Secondly, the preservative contains an acidifier which lowers the pH of the water. Water moves through a flower's vascular system best at a pH of 3.5. The acidifier also helps to reduce bacterial action.

Finally, preservatives also contain a **bactericide** which kills bacteria. Bacteria in the water clog the stems, restricting water uptake. They also cause the water to sour and smell bad. Additionally, controlling the growth of bacteria helps to reduce the production of ethylene gas.

A number of commercially prepared solutions are available. These can be purchased in liquid or powder form. The directions for mixing the preservatives should be followed exactly since too much preservative can burn the flower and too little will not be enough to keep the flowers fresh. It does not matter how deep the water and preservative solution is because flowers suck up water like a straw. An increase in depth cannot move more water up the stem.

If commercial preparations are not available, a preservative can be made at home. It is effective but requires more work and is expensive. Mix 50 percent warm water and 50 percent Sprite, 7-Up or similar drink containing citric acid. Add 1-1/2 teaspoons of household bleach to each quart of solution. The drink contains sugar, as a source of energy, and the citric acid in the drink is an acidifier. The bleach keeps bacteria from forming in the water.

6. *Allow flowers to absorb water.* All flowers with the exception of roses should remain in the warm preservative solution outside the refrigerator for 1 to 2 hours. Roses should be stored in the cooler immediately. This allows them to absorb a maximum amount of water. At the end of this time the flowers should feel turgid (full of water). Flowers that are shipped in bud such as gladioli, lilies, and carnations could sit at room temperature overnight to open up and reach optimum condition.

7. *Refrigerate.* The floral cooler becomes the storage place for flowers until they are passed on to the consumer (Figure 5-9). Even if flowers are to be used quickly they should first be cooled so that they become crisp and hard.

If the flowers become wet while being prepared, allow them to dry before being placed in the cooler. Wet flowers encourage the growth of botrytis.

The floral cooler provides a low temperature and high humidity. This combination helps to extend the life of the flowers. Low temperature slows down the activity of the flowers. The respiration and transpiration rates are lowered, as well as bacterial growth and the effects of ethylene gas. The high humidity in the cooler also slows down transpiration. This helps to prevent the flowers from drying out. Flowers last longer at 90 percent humidity.

FIGURE 5-9

A floral cooler is used to store flowers. *Photo courtesy of M. Dzamen*

The temperature at which the cooler should be set depends on the kind of flowers being stored. The ideal average temperature to keep most commonly used flowers, such as roses and carnations, is 36 to 40°F (2 to 5°C, or Celsius).

Tropical flowers, such as anthuriums and orchids, can be damaged at temperatures below 40°F (5°C). These flowers should be stored in a separate cooler at 45 to 50°F (7–10°C).

If only one cooler is available, the tropicals can be stored at room temperature. Select a cool area in the shop that is free of drafts.

Store flowers in the cooler in a manner that allows for good air circulation around the flowers. This will help to control diseases in the cooler. Clean the cooler regularly and keep it free of broken leaves and flowers.

ANTITRANSPIRANTS

The loss of water from flowers is known as transpiration, which was explained in the first section of this chapter. Antitranspirants such as Crowning Glory are materials that slow down the loss of water from the flower through transpiration.

Antitranspirants are usually diluted with water. Follow the directions on the label for mixing. The flowers can be dipped into the solution prior to being arranged or sprayed after the arrangement is completed. After applying, let the arrangement dry at room temperature before storing in the cooler.

Antitranspirants prevent browning of the petals and retard deterioration of the flower. The use of antitranspirants is one additional step florists can use to ensure the long life of their flowers.

SUMMARY

1. Always unpack flower shipments immediately.
2. Remove paper or plastic sleeves, and loosen flower bunches to provide air circulation and space for flowers to open. Do not remove sleeves on roses.
3. Remove damaged blooms and foliage.
4. Remove foliage below the water line.
5. Use a sharp knife to recut the ends of the stems under water.
6. Use floral preservatives.
7. Use nonmetallic containers.
8. Allow flowers to sit for one to two hours at room temperature.
9. Store flowers in well-ventilated areas at recommended temperatures.

10. Keep flowers away from ethylene gas sources (fruits, vegetables, and decaying flowers or foliage).
11. Keep all surfaces clean.
12. Spray flowers with an antitranspirant.

Student Activities

1. As flowers are received at your school for classroom use, properly condition and store them according to the guidelines presented in this chapter.
2. If your school has a refrigerator, use a thermometer and check the temperature inside the cooler.
3. Set up experiments to test the following:
   a. Flowers kept at low temperatures and high humidity last longer than flowers stored at room temperature.
   b. Preservatives extend the keeping quality of cut flowers.
   c. 7-Up and Sprite are effective floral preservatives.
   d. Ethylene gas has a negative effect on the keeping quality of cut flowers. This can be tested by storing some flowers in a plastic bag with ripened apples. Seal the bag, and store it in the cooler. In a few days, compare the flowers with other flowers stored in the cooler.

Self-Evaluation

A. Fill in the blank space with the answer that best completes each statement.

1. The process by which plants produce food in the leaves of the plant is called _____.
2. Plants lose water vapor from the plant by a process called _____.
3. _____ is produced by diseased or decaying flowers and foliage.
4. The stems of flowers should be cut with a sharp _____.

5. A material that is added to the water to extend the life of flowers is known as a floral _____.

6. A disease that causes brown spots on the petals of flowers is _____.

B. Matching

_____ 1. Chain of Life program

_____ 2. xylem

_____ 3. ethylene gas

_____ 4. "sleepiness" in carnations

_____ 5. a rag or heavy glove

a. water-conducting tubes in stems

b. developed by the Society of American Florists

c. ethylene gas

d. produced by fruit

e. used to remove leaves from the stems

C. Select the best answer from the choices offered to complete the statement.

1. Flowers such as roses and carnations should be stored at
   a. 45–50°F.
   b. 36–40°F.
   c. 30–35°F.
   d. 55–60°F.

2. Foliage below the water line should be removed to
   a. prevent the leaves from decaying, which fouls the water.
   b. prevent photosynthesis from taking place.
   c. prevent the leaves from using stored food in the stem.
   d. all of the above.

3. Flower stems should be recut, removing
   a. 1/2 inch to 1 inch of the stem.
   b. 2 to 3 inches of the stem.
   c. 1/2 of the stem.
   d. 1 to 2 inches of the stem.

4. Tropical flowers are best stored at
    a. 30–35° F.
    b. 35–40° F.
    c. 45–50° F.
    d. 60–70° F.
5. All flowers except roses should be allowed to remain in a warm preservative solution outside the refrigerator to
    a. adjust to their new environment.
    b. allow time for the flowers to lose ethylene gas.
    c. absorb a maximum amount of water.
    d. reduce bacterial growth.
6. Preservatives
    a. acidify the water.
    b. reduce bacterial growth.
    c. serve as a source of food.
    d. all of the above.
7. Floral containers should be washed using
    a. a hot detergent solution and bleach.
    b. a commercial cleaner.
    c. Windex.
    d. only clean water.

D. Short Answer Questions

1. Explain the significance of the Chain of Life symbol.
2. List the reasons for early flower deterioration.
3. What effect does water quality have on hydration?
4. What are the sources of ethylene gas?
5. Explain the term "conditioning."
6. Why should flower stems be cut at an angle?
7. List the functions of a flower preservative.
8. What is an antitranspirant?

Mechanics and Supplies Used in Floral Design

Terms to Know

anchor tape
container
floral clay
floral foam
floral preservative
florist shears
hot glue
mechanics
pick machine
picks
pruning shears
ribbon shears
*stem wrap or floral
    tape*
wire cutters

Materials List

adhesive clay
anchor tape
cutting tools
floral foam
floral tape

continued

OBJECTIVE

To identify the basic mechanics and supplies used in floral design.

Competencies to Be Developed

After completing this unit, you should be able to:

- select containers for floral designs.
- identify the different types of floral foam.
- prepare a block of floral foam for a design.
- identify adhesive materials used in floral design.
- select cutting tools to be used in floral design work.
- identify the types of picks used in floral design.

Introduction

Floral designers use a number of tools and materials to help them arrange flowers for vases and corsages. All of the materials used to assist the designer in placing and holding flowers are called **mechanics**. In this chapter, you will explore the use of some of the most commonly used mechanics. Later, as you begin making corsages and arranging flowers, you will learn to safely use each of these.

CONTAINERS

Anything that holds water can be used as a **container**. However, the container must help to express the idea the designer has in mind for the arrangement. If a designer selects delicate, white roses to create a formal feeling, a rustic clay container would not be as good a choice as a crystal one.

The container should add to the appearance of the arrangement and harmonize with the rest of the design and the display environment. For example, an arrangement in a silver container would not be appropriate for a backyard barbecue where it would not harmonize with its surroundings.

Containers are selected by their characteristics just as flowers are. When selecting a container, the designer should consider the texture, shape, size, and color of the container.

Texture

The texture of a container depends largely on the material of which it is constructed. Containers are generally made of glass, glazed pottery, plastic, wood, metal, and papier-mâché, which is made of paper pulp. The textures of each of these materials can vary according to their finish. Generally, we think of containers with a smooth finish as having fine texture and containers with a rough finish as having a coarse one. The texture of the container should be compatible with that of the flowers. Silver containers have a fine texture and require the use of such flowers as roses and lilies. The rough texture of a wooden basket, on the other hand, would suggest the use of coarsely textured flowers, such as daisies and zinnias.

Shape

The shape of a container is one of its most important characteristics. Commonly used shapes include vases, bowls, pedestals, and baskets (Figure 6-1). The shape of a container often determines the form of the design. Bowls, for example, are often used for horizontal designs and low table arrangements, while vases would be used for more vertical designs.

Containers should have clean lines and not be highly decorated. Highly decorated containers attract attention to themselves and away from the finished arrangement. Containers in

Materials List

continued
hot glue gun
a variety of different
   type containers
wood and steel picks

FIGURE 6-1

Some commonly used containers include vases, bowls, pedestals, and baskets.

the shape of clowns or automobiles, for example, should not be used except for special occasions.

Size

The principles of scale and proportion will help select a container of the appropriate size. The flowers for an arrangement should be 1-1/2 to 2 times the height of the container, and the overall arrangement should be in a good size relationship with its setting. An arrangement to be placed in a church would need to be large, therefore requiring a large container. An arrangement for a dining room buffet would be much smaller and utilize a proportionately smaller container. The size of the container must create a feeling of stability and harmony with the arrangement.

Color

The color of a container must harmonize with the color scheme of the flowers and the room where the flowers will be displayed. Container colors that do not blend with the colors of the flowers call attention to themselves and away from the flowers. A blending of colors is preferable to contrasting ones.

Green containers that echo the foliage are often used by florists. Other commonly used colors are tan, brown, gray, and white. Only use white containers if the flowers are mostly white. White can draw attention away from the focal point. The principle of unity requires that the color of the container and flowers harmonize.

FLORAL FOAM

The most commonly used material for holding the stems of flowers is **floral foam**. Chicken wire and needlepoint holders have limited use as holding devices, although before the manufacture of floral foam, these were used extensively (Figure 6-2). Today, chicken wire is used mainly along with floral foam to help support heavy stems in large arrangements. Needlepoint holders are sometimes used in designs where limited numbers of flowers are needed and floral foam would be visible.

Floral foam is a soft, very absorbent lightweight material. When the stems of flowers are placed in the foam, they are able to take up water through the foam. Floral foam can be purchased in cases of forty-eight blocks that are 3 inches by

FIGURE 6-2

Chicken wire and needlepoint holders have limited use in holding the stems of flowers.

FIGURE 6-3

Specialty forms and holding devices for floral foam.

4 inches by 9 inches. For specialty uses, foams enclosed in plastic cages are also available, as is foam for dried and silk flowers (Figure 6-3).

Floral foam may be purchased in several different densities. The Smithers-Oasis Company is the most widely known manufacturer of floral foam. This company is so well known that many florists refer to floral foam as oasis. The various kinds of floral foam produced by this company are listed in Figure 6-4.

Floral foam should be thoroughly soaked with water before use. Floral preservative should be added to the water. The manufacturer recommends that floral foam be soaked

TYPES OF FLORAL FOAM	USES
Standard	An all-purpose foam that is used with a large variety of flowers.
Deluxe	A sturdier foam used for heavy stemmed flowers.
Instant	An all-purpose foam that absorbs water quickly.
Instant Deluxe	A foam similar to deluxe but absorbs water quickly.
Springtime	A softer foam ideal for the soft stems of bulb crops.

FIGURE 6-4

Various types of floral foam.

FIGURE 6-5
Free-float method of
soaking floral foam.

using the free-float method (Figure 6-5). Place a block of floral foam into a bucket or sink filled with preservative-treated
water to a height greater than the foam. As the foam absorbs
the water, the block will gradually sink. When only 1/4 inch
of the block remains above the surface, the block is saturated and ready to be used. Instant foam absorbs water so
quickly, it can be placed in the container dry and water
poured over it when needed. Thoroughly soaking the floral
foam is most important. Dry spots in the foam can cause
flowers to wilt.

ADHESIVE MATERIALS

Four kinds of adhesive materials, as shown in Figure 6-6, are
used by most florists. They help to secure mechanics.

Waterproof or anchor tape is used to secure floral foam
into the container. Anchor tape is available in 1/2- and 1/4-
inch widths. The 1/4-inch width is preferred since it covers
less area when stretched over floral foam. The wider tape
covers more area and can cause problems in the placement
of flowers.

Anchor tape is available in green, white, and transparent
varieties. Use the white tape on white containers and the
green on colored containers. Green tape can be more easily
hidden in the floral arrangement than white.

Stem wrap or **floral tape** is a waxed, stretchy type material that only sticks to itself. The techniques for using this
material are explained in Unit 7. Stem wrap is mainly used

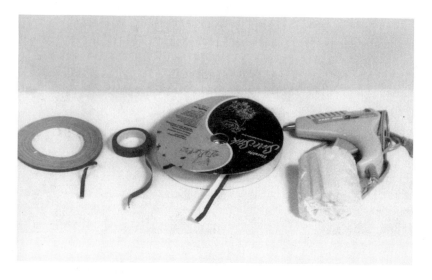

FIGURE 6-6

Four adhesive materials include waterproof tape, floral tape, adhesive clay, and hot glue.

in corsage work to create stems for the flowers used in the corsage. It is also sometimes used to hide wires or picks in other type designs.

Stem wrap is available in several widths, but 1/2 inch is the most commonly used. It is sold in an assortment of colors to match the colors of the design. The most frequently used colors are green and white.

Floral adhesive clay is a sticky type material similar in texture to children's play dough. It comes in a flattened strip wound onto a roll, with waxed paper between the layers. Two commonly sold brands are Cling and Sure-Stik.

Floral clay is used to fasten anchor pins (round plastic holders with four upright prongs used to hold floral foam in place) or to anchor pinpoint holders (Figure 6-7). Pinpoint holders are round devices of steel needles used to hold flowers. Small pieces of the clay are placed on the bottom of the holder, which is then pressed firmly against the container. Make certain the container surface is dry. Floral clay will not stick to a wet surface. Floral clay has one disadvantage. It will leave a sticky mark on the surface of the container which is difficult to remove. As a designer, you may want to be cautious in the use of clay on expensive containers.

Hot glue is an adhesive material used extensively in the florist shop. It is purchased in solid sticks that are inserted into an electric gun to be melted into a liquid that hardens within seconds. Hot glue can also be melted in a glue pot or

FIGURE 6-7

Use adhesive clay to fasten anchor pins and pinpoint holders in place.

electric frying pan. A low-temperature hot glue is also available. It does not hold as well as the regular hot glue but is safer to use. Hot glue has a wide range of uses that includes design work with live and silk flowers.

Caution: Hot glue guns and hot glue can severely burn your skin.

CUTTING TOOLS

A variety of cutting tools are used by florists in the construction of corsages and arrangements (Figure 6-8). Each of these has a specific use. You will find that tools last longer if they are used correctly and properly maintained.

Floral knives are one of the most important tools used by the designer. The knife is used for cutting stems only.

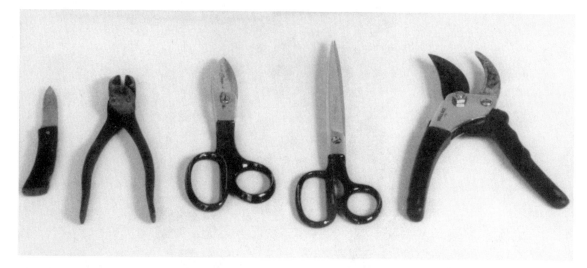

FIGURE 6-8

Cutting tools used by the florist include a floral knife, wire cutters, florist shears, ribbon shears, and pruning shears.

Never attempt to cut wires with a knife. You may be successful in cutting small wires, but this will dull your knife.

Most florists prefer a knife for cutting stems because they make a quick clean cut. Other tools, such as shears, will cut the stem but may pinch the tubes that take up water, thereby restricting the flow of water to the plant.

Select a knife with a short, sharp blade that can be easily handled while working. Most designers hold the knife in their hand when not actually cutting. Laying the knife down after each cut would be a waste of time.

Keep your knife sharp. You are less likely to cut yourself with a sharp knife. Hold the knife between the thumb and forefinger in the manner illustrated in Figure 6-9. Cut toward yourself at an angle. Move your entire hand including your forefinger so that you do not cut yourself. Do not attempt to cut hard woody stems in this manner. Ask your instructor to illustrate the correct use of the knife before you begin. Proceed slowly and with caution until you feel comfortable using the knife.

Florist shears are another tool that can be used for cutting stems. However, this tool is slower to use since it has to be laid down after each cut and as discussed earlier, it tends to pinch the stems restricting the uptake of water. The primary uses of florist shears are cutting wires and ribbons.

FIGURE 6-9
Making a knife cut.

Ribbon shears or scissors are useful for cutting ribbons and decorative foils used by the florist. Never cut stems or wires with ribbon shears.

Wire cutters are used for cutting wires and the stems of artificial flowers which contain a wire. Use wire cutters for cutting wires only, not on plant material or ribbon.

Pruning shears are useful for cutting heavy stems that are too large to be easily cut by a knife or florist shears. These are particularly useful when cutting woody stemmed materials that are sometimes used in arrangements. Pruning shears may be used to cut stems of up to 1/2 inch in diameter.

PICKS

Florists use wooden and steel **picks**. These are used mainly in funeral designs, to make wreaths, and in dried and artificial arrangements. Wooden picks can be purchased in the natural wood color or the more common green color (Figure 6-10). They come with or without a wire attached and may be from 2-1/2 to 7 inches in length. Wooden picks are used to add length or support the plant materials.

Steel picks are commonly sold in lengths of 1-3/4, 2-1/8, and 3 inches. They are attached to the stems of various materials with a steel **pick machine** (Figure 6-11). The pick

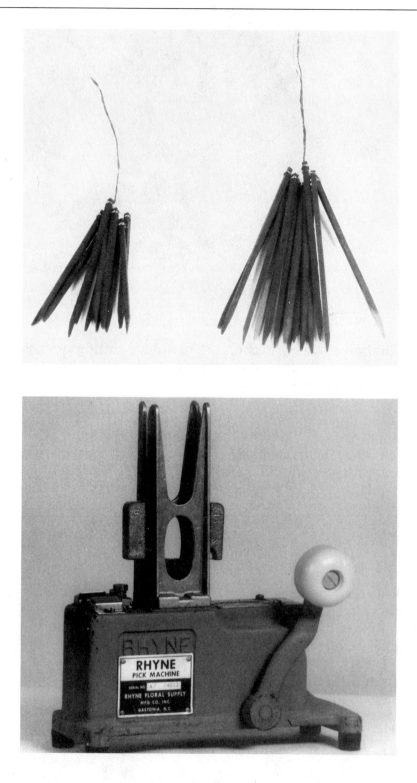

FIGURE 6-10
Wooden picks.

FIGURE 6-11
Steel pick machine.

makes it easy to insert plant material into a wreath ring, Styrofoam, or other holding device.

Caution: Steel picks are sharp and may cut your fingers.

The mechanics discussed in this chapter represent only a small portion of the mechanics used by florists. You will be learning about others in subsequent chapters.

Student Activities

1. Set up a display of various containers and share with the rest of the class the characteristics of the containers and what type of arrangement would be best suited for each container.
2. Ask your instructor to demonstrate the use of a knife for cutting stems. Practice cutting plant material until you feel comfortable using a knife.
3. Examine the cutting tools you have in your classroom and explain the use of each.

Self-Evaluation

A. Select the answer from the choices offered that best completes each of the statements.

1. The container most suited for white roses is
   a. an earthenware bowl.
   b. a basket.
   c. a crystal vase.
   d. a ceramic figurine bowl.
2. A size of a container selected for a design is determined by
   a. the size of the flowers to be used.
   b. the place where the arrangement is to be displayed.
   c. the shape of the design.
   d. both a and b.

3. The shape of a container
   a. often determines the form of the design.
   b. determines the stability of an arrangement.
   c. has little to do with the shape of the arrangement.
   d. determines the value of the arrangement.
4. The most commonly used material for holding flowers is
   a. pin-point holders.
   b. frogs.
   c. chicken wire.
   d. floral foam.
5. Floral foam should be saturated with water using the
   a. drenching method.
   b. free-float method.
   c. container as a holding device.
   d. water hose.
6. Waterproof tape is used to
   a. tape the stems of corsages.
   b. tape stems to the sides of the container.
   c. anchor the heads of flowers.
   d. secure floral foam in the container.
7. The most commonly used width of stem wrap is
   a. 3/4 inch.
   b. 1 inch.
   c. 1/2 inch.
   d. 1/4 inch.
8. The knife used by floral design students should
   a. be long-bladed and sharp.
   b. be short-bladed and sharp.
   c. be long-bladed and dull so you won't cut yourself.
   d. be short-bladed and dull so you won't cut yourself.

9. Florist shears may be used to cut
    a. stems, ribbon, and wires.
    b. stems and ribbon only.
    c. stems only.
    d. wires and stems.
10. Deluxe floral foam is used to
    a. hold soft-stemmed flowers.
    b. absorb water quickly.
    c. act as a preservative.
    d. hold heavy-stemmed flowers.
11. Wooden picks
    a. may be purchased with or without a wire.
    b. may be purchased in lengths of up to seven inches.
    c. may be purchased in natural wood or green color.
    d. all of the above.

B. Short Answer Questions

1. What type of container would be appropriate for a formal party? Explain why.
2. How does texture affect the selection of containers?
3. What is the recommended procedure for soaking floral foam?
4. Why is a knife better than shears for cutting stems when making an arrangement?

Selecting Wire and Wiring Flowers

Terms to Know

calyx
gauge
petiole
taping

Materials List

cut flowers of different
*   flower types*
floral knife
floral shears
floral tape
20-gauge wire
24-gauge wire
26-gauge wire
28-gauge wire
wire cutters

OBJECTIVE

To wire flowers for corsages, arrangements, and wreaths.

Competencies to Be Developed

After completing this unit, you should be able to:

- select wire of appropriate size for specific flowers and foliage.
- select the correct wiring procedure for different flowers and foliage.
- wire a given flower using the correct wiring method.
- tape a flower stem or wire using florist tape.

Introduction

Wiring techniques are essential to the floral designer and are a basic skill you must master as a student in floral design. Florists will not wire a flower if the flower can be used satisfactorily without it. However, many cases require the use of wire.

The florist wire is used for a number of reasons to:

1. straighten slightly crooked stems.
2. support weakened stems.
3. keep flowers upright and help prevent wilting.

4. hold flowers and foliage in a desired position.
5. prevent flower heads from breaking off the stem.
6. replace flower stems on corsages so the corsage stem is not bulky.
7. add accessories to corsages and arrangements.

SELECTING WIRES

Florist wires are commonly sold in 12-pound boxes containing straight wires 18 inches in length. The number of wires per box varies according to the size of the wire (Figure 7-1). The wires are coated with green enamel. The enamel coating helps prevent the wires from rusting. The green color makes them less noticeable in the floral design. Wires may also be purchased on spools for special needs such as making garlands (wreaths of flowers or foliage).

SIZES AND USES OF WIRE

Florist wire comes in various weights and diameters called **gauges**, which range from sizes 18 (thickest) to 32 (thinnest). Wire gauge numbers decrease as the wire gets larger. If you are limited in the variety of wire gauges available, select 20- and 26-gauge wire. This will allow you to complete most projects needed in a floral design class.

WIRE GAUGES	NUMBER/BOX
# 16 wire	768
# 18 wire	1,320
# 19 wire	1,776
# 20 wire	2,472
# 21 wire	2,976
# 22 wire	3,660
# 23 wire	4,500
# 24 wire	5,664
# 26 wire	9,156
# 28 wire	11,432
# 30 wire	15,304

FIGURE 7-1

Wire sizes and number of wires per box.

The proper selection of the correct size wire is important to the floral design student. Use too large a wire, and the flower may be damaged. If a wire is too thin, the flower will not be properly supported. The goal is to select the smallest wire that will support the flower and still hold it in place. Wiring should only be done when necessary. Using too much wire can detract from the arrangement.

Figure 7-2 lists some of the most commonly used flowers and a range of recommended wire sizes. The wire size you choose depends on the weight of the flower and the intended use.

METHODS OF WIRING FLOWERS

There are a number of ways to wire flowers depending on the flower type and the use of the flower. Six methods are presented in this text.

Caution: Wires can easily puncture your finger. Never hold your finger over the area where a wire is to be inserted or where it will exit the flower.

FLOWER	WIRE SIZE
Rose	20–22
Carnation	20–22
Standard Chrysanthemums	18–20
Pompon Chrysanthemums	20–24
Gladiolus	18–20
Snapdragons	20–22
Calla Lily	18–20
Stephanotis	26–28
Tulips	20–22
Orchids	24–28
Stock	18–20
Gerbera Daisy	20–22
Freesia	22–24

FIGURE 7-2

Recommended wire sizes for commonly used flowers.

FIGURE 7-3

The straight wire method of wiring flowers.

Straight-Wire Method

The **straight-wire method** is used when the stem remains attached to the flower. This method is most often used when wiring flowers for vase arrangements (Figure 7-3).

1. Hold a piece of 20-gauge wire about 1/2 inch from the end, and insert it into the **calyx**, the fleshy part of the flower below the petals. Push it up toward your finger.
2. Wrap the wire carefully around the stem going between the leaves. Remember, you want the wire to show as little as possible.

Hook Method

The **hook method** is used on daisies, asters, chrysanthemums and other flowers used for corsages, funeral work, or arrangements. This method is recommended for any flower that easily breaks off at the stem. The hook used in this method prevents that from happening. The stem of the flower may be removed or left intact depending on the intended use of the flower (Figure 7-4).

1. If the flower is to be used in a corsage, cut the stem 1/2 inch below the calyx. Remove any foliage remaining on the stem.
2. Hold an appropriately sized wire near the end and push it up through the calyx and out the top of the flower.

FIGURE 7-4
The hook method.

This wire may also be pushed up along the center of the flower stem.

3. Bend the end of the wire into a small hook.
4. While holding the flower so that the head is supported, pull the wire downward so that the hook disappears into the flower head.
5. If the flower is to be used in a corsage, it is ready for **taping** which is discussed later in this chapter.
6. If the flower is to be used in an arrangement, bend the wire around the stem as explained in the straight-wire method.

Piercing Method

The **piercing method** of wiring is used on flowers having an enlarged calyx, such as carnations and roses (Figure 7-5).

1. Remove the stem of the flower about 1/2 inch below the calyx if the flower is to be used in a corsage.

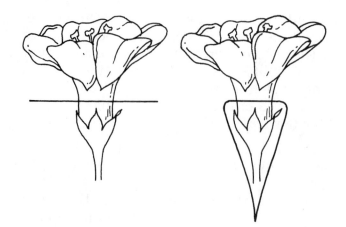

FIGURE 7-5
The piercing method.

2. Insert a wire through the calyx at a point halfway between the base of the calyx and the petals. If the flower is to be used in a vase arrangement, push the wire until it extends about 2 inches beyond the calyx, then bend both sides of the wire down and wrap the longer end of the wire around the stem. If the finished product is to be a corsage, push the wire through the calyx to the midpoint of the wire and then bend both sides of the wire down. If additional support is needed, a second wire may be inserted at a 90° angle to the first and bent down in similar fashion.
3. Tape the flower if it is to be used in a corsage.

Hairpin Method

The **hairpin method** is used on tube-shaped flowers such as stephanotis and certain lilies (Figure 7-6).

1. Cut the stem 1/2 inch below the calyx.
2. Bend a thin wire, such as 26-gauge wire, at its midpoint, creating a hairpin shape.
3. Place a small piece of moist cotton into the center of the hairpin.
4. Insert the ends of the hairpin through the top of the flower tube so that it exits at its base.
5. Gently pull the wires until the cotton wad is held firmly in place against the inside of the flower. The wires will rest along the sides of the stem.
6. Tape the flower starting at the base of the flower and working down.

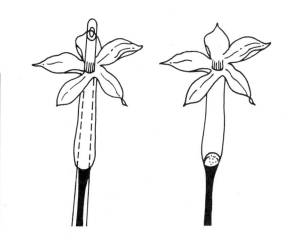

FIGURE 7-6

The hairpin method.

FIGURE 7-7

The wrap method of wiring flowers and foliage.

Wrap Method

The **wrap method** is used on foliage made of many small leaflets, such as leatherleaf or flowers composed of many small florets (Figure 7-7).

1. Cut a stem of leatherleaf so that a small section of stem remains on the leaf.
2. Make a hairpin from a 26- or 28-gauge wire.
3. Hang the hairpin over the lowest pair of leaflets so that the bend in the hairpin rests against the stem.
4. Wind one wire over both the other wire and the stem of the leaflet.
5. Wrap the stem and wire with floral tape.

Stitch Method

The **stitch method** of wiring is used mostly on broad, leathery-skinned leaves, such as camellia, pittosporum, and salal (Figure 7-8). Foliages that have been wired are much easier to use because the wire creates an extended **petiole**, or leaf stem. This makes it easier to place and shape the leaf in the corsage.

Working from the back side of the leaf:

1. Take a 26-gauge wire and, while slightly bending the leaf lengthwise, pass the wire through the leaf under

the mid-vein. This stitch should be made about midway up the leaf.

2. Bend both ends of the wire down along the back of the leaf.
3. Wrap one half of the wire around the other half and the petiole several times.
4. Tape the stem beginning at the base of the leaf.

The wiring techniques presented in this chapter are the most commonly used methods in the flower shop. Other specialized wiring techniques are also used on fragile or uniquely shaped flowers. The phalaenopsis orchid is an example of such a flower.

When selecting a wiring method for a flower, consider the type of stem and the flower head involved. Also consider whether the flower is to be used in a corsage or in a vase arrangement. For example, a rose to be used in an arrangement could be wired with the straight wire method. If the rose is to be used in a corsage, it would probably be wired using the piercing method. Keep all wire hidden as much as possible. Unsightly wires detract from the beauty of the corsage or arrangement.

FLORAL TAPING

Florist tape is a nonsticky tape that will stick to itself when stretched. It is used mainly in corsage work to cover wires, bind wires to flower stems, and to bind wired and taped flowers together.

Florist tape is available in 1/2-inch and 1-inch widths. The 1/2-inch width is most commonly used and you may have difficulty finding the 1-inch size.

Florist tape is sold under the brand names Floratape and Parafilm. It is available in a variety of colors. Foliage green and moss green are the most frequently used. White is often used for wedding work and brown is often used to wrap stems of dried flowers. Other colors available include black, lavender, red, yellow, blue, and orange.

Taping flowers and wires with florist tape requires skill, which develops with practice. Use the following steps as a guide. These are given for a right-handed person. Reverse the hands if you are left-handed.

FIGURE 7-8

The stitch method of wiring foliage.

Taping a Wire

1. Hold an 18- or 20-gauge wire in your left hand near the top.
2. Place a roll of Floratape over your little finger and guide the tape across the palm of your hand and hold the tape between the thumb and index finger. If this feels awkward, then unroll about two feet of tape and place the roll on a table. Guide the tape across the palm of your hand and hold it with your thumb and index finger. Many florists use the bars of a steel pick machine to hold their floral tape.
3. Stretch the end of the tape to activate the stickiness of the tape. Place the end of the tape under the wire; roll a small amount around the wire and pinch it so that the tape sticks to the wire (Figure 7-9).
4. Begin rotating the wire in your left hand between your fingers and thumb.

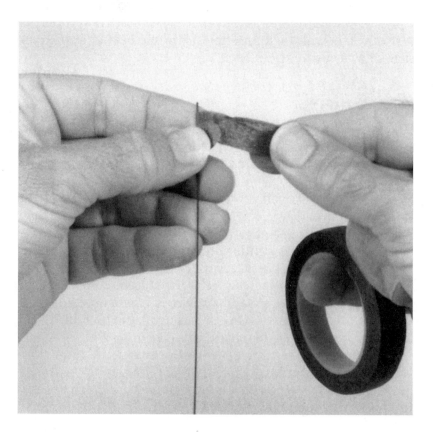

FIGURE 7-9

Attaching florist tape to a wire.

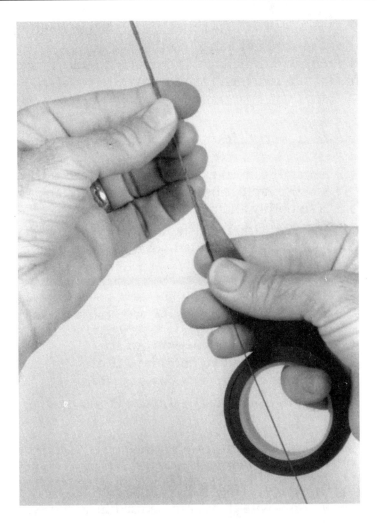

FIGURE 7-10

The angle at which florist tape is held determines the amount of overlap.

5. Pull downward with the tape in your right hand while you guide the wire with your fingers. The tape should be held at a sharp angle (20 to 30°) to the wire (Figure 7-10). The angle determines the amount of overlap. If the angle is greater than 30°, the tape of the wire will be thick and heavy. A sharp angle allows the wire to be covered with minimum overlap.

6. Continue rotating the wire and maintaining tension on the tape until you reach the bottom of the wire. Pull sharply on the tape and it will break away from the wire.

A properly wrapped wire is smooth and lightly taped. If the tape does not adhere to the wire and is loose, increase the tension on the tape. If the tape is bulky in spots, then hold it at a consistently sharp angle.

Student Activities

1. Select 18- or 20-gauge wires and a roll of florist tape. Practice taping the wires until you can tape a wire that is smoothly and uniformly covered.
2. Practice wiring and taping flowers and foliage using each of the wiring methods discussed in this chapter.

Self-Evaluation

A. Select the best answer from the choices offered to complete the statement or answer the question.

1. Most florists
   a. always wire flowers.
   b. will not wire flowers if the flowers can be used satisfactorily without it.
   c. do not recommend wiring flowers.
   d. only wire flowers to be used in expensive arrangements.
2. Florist wires are used
   a. to support weakened stem flowers.
   b. to prevent flower heads from breaking off.
   c. to hold flowers and foliage in a desired position.
   d. all of the above.
3. The length of florist wires is
   a. 24 inches.
   b. 36 inches.
   c. 12 inches.
   d. 18 inches.

4. The larger the gauge of florist wire, the
   a. smaller the wire.
   b. larger the wire.
   c. longer the wire.
   d. shorter the wire.

5. To wire a rose for a vase arrangement, use a
   a. 26-gauge wire.
   b. 32-gauge wire.
   c. 20-gauge wire.
   d. 28-gauge wire.

6. When taping a wire, the _____ determines how tight or loose the tape is.
   a. tension
   b. angle
   c. hook
   d. elasticity

B. Match the following flowers and foliages with the wiring method best suited for each flower or foliage.

_____ a. rose (for a vase arrangement)

_____ b. rose (for a corsage)

_____ c. carnation

_____ d. leatherleaf

_____ e. chrysanthemum

_____ f. daisy

_____ g. camellia leaf

_____ h. stephanotis

1. hook method
2. stitch method
3. wrap method
4. straight-wire method
5. piercing method
6. hairpin method

C. Short Answer Questions

1. List the common uses for floral tape.
2. How does a florist determine if a flower needs wiring?
3. List the reasons for using florist wire.
4. How does a florist determine which size wire to use when wiring a flower?

Selecting Ribbons and Tying Bows

Terms to Know

accessory
bolts
double-faced
satin
single-faced
texture

Materials List

one bolt #3 satin
ribbon
one bolt #9 satin
ribbon
one bolt #40 satin
ribbon
scissors
26-gauge florist wire
or chenille stems
wire cutters

OBJECTIVE

To tie attractive bows for corsages, wreaths, and flower arrangements.

Competencies to Be Developed

After completing this unit, you should be able to:

- select ribbon of appropriate size for a specific design.
- identify materials from which ribbons are made.
- construct a florist bow from several sizes of ribbon.

Introduction

As a beginning designer, you need to be knowledgeable about ribbons and skilled in the tying of bows. The bow is an important **accessory** (a nonessential addition) in many floral arrangements. It is used to give the design a finished look and can be used to create a mood or feeling. For example, red and green plaid ribbon makes a person think of Christmas while red, white, and blue colors are associated with the Fourth of July. Ribbons can also be used to make a design look more expensive. While bows are not difficult to tie, much practice is necessary to achieve proficiency.

SELECTING RIBBONS

The ribbons used in flower shops today vary greatly in materials, textures, and patterns, with many novelty ribbons being introduced each year. These ribbons give the designs a different look and catch the eye of the consumer. Ribbons come in an assortment of solid colors, prints, plaids, and many other combinations. Because of these differences, some ribbons are more appropriate for certain occasions than others.

Ribbon is either **single-faced** or **double-faced.** Most ribbon is single-faced and has a shiny side and a dull side. Double-faced ribbon has the same finish on both sides.

Today's designers have a wide selection of ribbon materials from which to choose: cotton, velvet, satin, nylon, acetate, plastic, burlap, paper twist, and lace (Figure 8-1). Curling ribbon, another popular ribbon used in flower shops, is made of polypropylene.

Satin is made of a glossy fabric and is probably the most frequently used ribbon in the flower shop. Satin is the easiest ribbon with which to work, and it comes in a wide selection of designs, colors, and widths (Figure 8-2).

Texture is another important consideration when selecting ribbons. Texture refers to the surface appearance of the ribbon and varies from fine to coarse. Satin ribbon has a fine texture, cotton, a medium one, and burlap the coarsest. For formal occasions, the designer will select finely textured ribbons. Ribbons used for informal occasions, such as an outdoor barbecue, would be coarse.

Ribbon Sizes

Ribbon is sold on cardboard spools called **bolts** and comes in a variety of sizes. Ribbon size follows an industry standard illustrated in Figure 8-3. The most common ribbon sizes are numbers 1-1/2, 3, 9, and 40. Some ribbon manufacturers no longer make number 100 or number 120 ribbon, and these sizes may be difficult to find at your local florist. The number 1-1/2 and number 3 ribbons are used primarily for corsages. Number 9 ribbon is mainly for decorating flower pots, and number 40 ribbon, for wreaths and sympathy flowers.

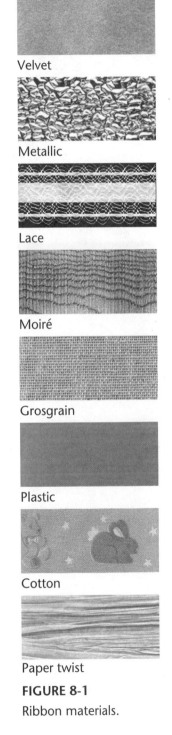

Velvet

Metallic

Lace

Moiré

Grosgrain

Plastic

Cotton

Paper twist

FIGURE 8-1

Ribbon materials.

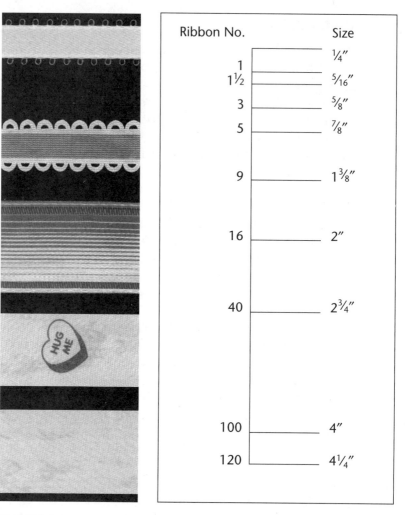

Ribbon No.	Size
1	¼″
1½	⁵⁄₁₆″
3	⅝″
5	⅞″
9	1⅜″
16	2″
40	2¾″
100	4″
120	4¼″

FIGURE 8-2

Satin ribbon.

FIGURE 8-3

Ribbon sizes and the numbering system.

TYING A BOW

The size of a bow should be in proportion to the size of the design. The ribbons of the corsage, for example, should extend just beyond the petals of the flowers. The bow should accent, not draw attention from, the flowers.

To become skilled at making bows, you will need lots of practice. Make several bows with number 3 ribbon using the following procedure. When you are satisfied with the quality of your bow, repeat the process using number 9 and number 40 ribbon. Because these ribbons are wider, you will need to increase the size of your bow. Depending upon the bow's intended use, you may want to increase the number of loops also.

There are many methods of tying bows. The one presented here is the most popular method used by florists for decorating corsages. To tie a corsage bow follow these directions.

1. Cut a 1-1/2-yard long piece of single-faced, number 3 satin ribbon. You will also need a 9-inch piece of 26-gauge wrapped wire or a chenille stem.
2. Wrap the end of the ribbon over your thumb forming a loose loop slightly larger than your thumb. The shiny side of the ribbon should face out. Hold the ribbon between the thumb and forefinger (Figure 8-4). Pinch it so that it narrows beneath the thumb.
3. The unused portion of the ribbon will have the dull side showing. Using the thumb and forefinger of your free hand, reach under the thumb and make a half turn in the ribbon so its shiny side shows. Hold the ribbon tightly to prevent slipping.
4. Turn the unused portion of the ribbon under and make a 2-inch loop. Lift the thumb slightly and slip the ribbon between the thumb and forefinger. Pinch the ribbon

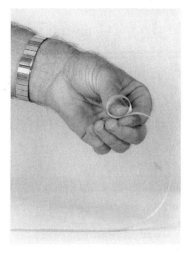

FIGURE 8-4
Form a loop over the thumb.

FIGURE 8-5
Make a 2-inch loop.

FIGURE 8-6
Make a second 2-inch loop.

again. This will form the first loop of the bow (Figure 8-5).

5. The dull side of the ribbon will be showing again. Make a half turn of the ribbon using the procedure in step 3.

6. Form an identical loop on the opposite side of the first loop using the same procedure (Figure 8-6).

7. Add three additional loops to the left and right side of the bow. Each loop should be slightly larger than the previous loop (Figure 8-7). Using scissors, cut the ribbon at an angle so that a 3- to 4-inch streamer remains.

8. Cut a piece of ribbon 6 to 8 inches in length. Holding the length of ribbon in the middle, add it to the loops between your thumb and forefinger (Figure 8-8). Pinch the ribbon.

9. Tie the bow with a taped wire. The wire is run through the loop formed over the thumb. Both ends of the wire are pulled to the back of the bow. Pinch the two wires together and twist tightly, as close to the ribbon as possible. The bow must be tied tightly or the loops will slip and not hold their position.

10. If the bow appears poorly shaped, place the bow in your hand with the thumb in the loop and your fingers beneath the bow as shown in Figure 8-9. Holding the bow

FIGURE 8-7

Add three additional loops on each side of the bow.

FIGURE 8-8

Add a piece of ribbon 6 to 8 inches long to form streamers.

FIGURE 8-9

Pull loops into place to shape the bow.

FIGURE 8-10
The completed bow.

tightly, pull the loops with the other hand and separate them until the bow appears as illustrated in Figure 8-10.

Do not be discouraged if your first bow does not look as perfect as you would like. With practice, your bows will improve.

Student Activities

1. Construct a bow from numbers 3, 9, and 40 ribbon and ask your teacher to evaluate the bows as you finish each one. To save ribbon while you are practicing, untie the bows and iron the ribbon with a cool iron.

2. Visit a local florist shop and study the kinds of ribbon used by the florist. Ask the staff to show you the store's procedure for making bows.

3. Make a display of various ribbon sizes and materials.

Self-Evaluation

A. Multiple Choice

1. The bow is not the focal point of a corsage and is considered to be

    a. an essential element.

    b. an accessory.

    c. of little importance.

    d. none of the above.

2. Single-faced ribbon

    a. is dull on both sides.

    b. is shiny on both sides.

    c. is shiny on one side and dull on the other side.

    d. is not available at your local florist.

3. Some of the materials from which ribbons are made include

    a. cotton, satin, linen.

    b. satin, corduroy, velvet.

    c. satin, cotton, velvet.

    d. all of the above.

4. The ribbon most frequently used by florists is made of

    a. velvet.

    b. cotton.

    c. paper.

    d. satin.

5. The surface appearance of a ribbon is called the

    a. balance.

    b. weight.

    c. bias.

    d. texture.

6. The ribbon size most suitable for a carnation corsage is

    a. number 3.

    b. number 9.

    c. number 40.

    d. number 100.

7. The ribbon size most suitable for decorating a pot mum is

    a. number 3.

    b. number 9.

    c. number 40.

    d. number 100.

8. The ribbon size most suitable for decorating a 16-inch wreath is
   a. number 1-1/2.
   b. number 3.
   c. number 9.
   d. number 40.
9. The size of the bow for a corsage should be
   a. four inches.
   b. as wide as the corsage is long.
   c. one-half the length of the corsage.
   d. slightly wider than the petals of the corsage.

B. Short Answer Questions

1. List five different materials used to make ribbon.
2. How does ribbon contribute to creating a certain mood or feeling?
3. How does texture affect your choice of a ribbon?

Boutonnieres and Corsages

boutonniere
corsage
floral adhesive
hot glue
net tufts
tulle

OBJECTIVE

To design boutonnieres and corsages.

Competencies to Be Developed

After completing this unit, you should be able to:

- design a single or multiple flower boutonniere.
- design a single-bloom carnation corsage.
- design a single-spray mini-carnation corsage.
- design a double-spray mini-carnation corsage.
- design a corsage using net tufts and artificial leaves.
- design a football chrysanthemum corsage.
- design a carnation corsage using hot glue.

Introduction

Flowers are worn by both men and women on special occasions such as weddings, proms, and holiday celebrations. Flowers worn by women are called **corsages**. A corsage is a cluster of flowers, foliage, and accessories that accents a woman's dress and adds to the theme of the celebration.

Flowers worn by men are called **boutonnieres**. A boutonniere, pinned to the man's lapel, usually consists of a single flower with foliage. There has been a recent trend toward more elaborate boutonnieres consisting of multiple flowers, such as two, or even three, stephanotis blossoms with foliage. The boutonniere adds a touch of color to the man's clothing and is usually coordinated with the theme of the occasion. If it is to be worn to a prom, its flowers should complement the flowers worn or carried by the man's date. For a wedding, the flowers would be selected to echo the overall theme of the event.

BOUTONNIERES

Select flowers for a boutonniere that have been properly conditioned (see Unit 5) and that hold up well out of water. The most popular flowers for boutonnieres are roses and carnations. Other flowers often used are stephanotis, pompon chrysanthemums, lilies of the valley, and alstromeria lilies.

Boutonnieres may be designed in a number of ways, one of which is illustrated in this text.

Constructing a Single-Bloom Carnation Boutonniere

Step 1.   Select materials:

> 1 standard carnation
> 1 small piece of leatherleaf fern
> 1 strand of 24- or 26-gauge wire
> florist tape
> boutonniere pin and bag

Step 2.   Cut the stem of the carnation just below the calyx, and wire the carnation using one half of a 24- or 26-gauge wire. Use the piercing method (refer to Unit 7).

Step 3.   Tape the stem using green floral tape. Begin taping the carnation above the wire high on the calyx of the carnation (Figure 9-1).

Step 4.   Select a tip of leatherleaf fern in proportion to the size of the carnation so that just the tips of the fern extend beyond the carnation. The fern may be wired using the wraparound method; however, many

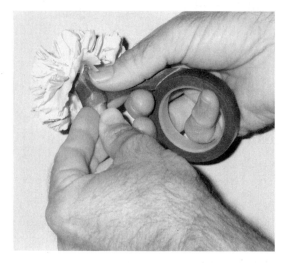

FIGURE 9-1
Tape the wired carnation.

FIGURE 9-2
Tape a sprig of foliage to the back of the carnation.

FIGURE 9-3
Stem treatment for boutonnieres.

florists skip this step when using leatherleaf fern. Strip the foliage from the bottom inch of the stem.

Step 5. Place the fern behind the carnation so that the tips of the fern are exposed. Use floral tape to bind the foliage and flower together (Figure 9-2). Use only enough tape to cover the stem smoothly. You do not want the stem to become bulky.

Step 6. Cut the stem to approximately 2 inches in length. Wrap any exposed wire. The stem may be left straight, curved, or bent to form a small hook at the base (Figure 9-3).

Step 7. Mist the flower with water and place in a boutonniere bag. Close the bag with a boutonniere pin that later can be used to pin the boutonniere (Figure 9-4). Refrigerate.

Once you've mastered this basic boutonniere construction, you can vary it using different flowers and multiple blooms (Figure 9-5). Use the criteria in Figure 9-6 to evaluate the boutonnieres.

FIGURE 9-4
Place the completed boutonniere in a bag.

FIGURE 9-5
Different styles of boutonnieres.

CRITERIA	YES	NO
1. The flowers are wired correctly and tape is smooth and free of wrinkles.	☐	☐
2. Foliage is appropriately sized for flowers and is firmly attached.	☐	☐
3. Bow is of correct size for the corsage and is firmly attached.	☐	☐
4. All wires are wrapped smoothly.	☐	☐
5. Corsage or boutonniere is packaged correctly.	☐	☐
6. Flowers have not been damaged.	☐	☐
7. Corsage or boutonniere is saleable.	☐	☐

FIGURE 9-6

Evaluating boutonnieres and corsages.

CORSAGES

The principles of design apply to corsages as well as to arrangements (refer to Unit 3). Use these principles as a guide as you learn to construct corsages.

Corsages are commonly worn on the lady's left shoulder. However, if the woman's dress is strapless or if the dress is made of a material that will not support the corsage, the corsage may be modified to be worn on the wrist, in the hair, or on her purse. Band aids may be glued to the bottom of the corsage and taped to the shoulder (see color insert).

The flowers and colors used for the corsage are dictated by the personal preferences of the wearer, the formality of the occasion, the size of the wearer, and the color and style of the clothing to be worn.

Florists should make an effort to identify the personal preferences of the individual customer. Find out if certain flowers are preferred or disliked. It does not matter how well a corsage is designed if the flowers displease the wearer.

The florist should also find out the occasion for which the corsage is prepared. Knowing the formality of the event helps the designer select flowers and accessories. A formal occasion would call for more expensive flowers with higher quality accessories. The design would also be more stylish.

The size of the wearer is another factor to consider when selecting flowers and styling the design. A small woman looks best with a small corsage, while a large woman needs larger flowers or a larger design. For example, a large cattleya orchid corsage would be unsuitable for a small woman. Likewise, a small carnation corsage would not look appropriate on a large woman.

The designer also needs to know the color of the garment the person will be wearing. The dominant color of the flower should complement the clothing. Use a color wheel to help you select pleasing hues. If the clothing color is not known, select white flowers and accessories. White flowers are considered appropriate for any occasion.

The corsage should be designed so that it is light and easy to wear. A woman does not feel comfortable wearing a corsage that is heavy and pulls on her dress. Removing the stems from flowers helps to keep them light. Also, space the flowers so that each can be seen separately. The designer does not want to make the corsage look skimpy, but the flowers should not be tightly concentrated either.

One of the most important factors to consider in designing flowers is good workmanship and good quality flowers. All wires should be covered. This is crucial to the professional appearance of the corsage and also prevents the wearer from being pricked by an uncovered wire. Good

| S-Curve | Straight Line | Round | Triangular | Crescent |

FIGURE 9-7
Corsage designs.

workmanship or mechanics means that the corsage will not come apart or be misshapen.

Corsage Design

Once the colors and flowers have been selected, the designer may choose from a number of styles or designs of corsages (Figure 9-7). Which style the designer selects will depend upon the kind of flowers being used and the personal preference of the wearer.

Constructing a Single-Bloom Carnation Corsage

Step 1.   Select materials:

> 1 standard carnation
> 1 stem of leatherleaf fern
> 1-1/2 yards of number 3 satin ribbon
> 1 24- or 26-gauge wire
> stem wrap or floral tape
> corsage bag and pins

Step 2.   Using the piercing method, wire and tape a standard carnation using half the length of a 24-gauge wire. This step is the same as that used in wiring and taping boutonnieres.

Step 3. Select a piece of leatherleaf fern for the backing of the corsage. Sprengeri fern or other foliage may be substituted. A tip 3 to 4 inches long will be needed. Strip the foliage off the bottom inch of the stem. This foliage may or may not be wired depending upon the designer. Beginning just below the calyx of the carnation, tape the foliage securely to the back of the carnation (Figure 9-8).

Step 4. Make a bow using number 3 or 1-1/2 satin of the same width as the fullest part of the carnation (refer to Unit 8). Tie it so that it is in proportion to the flower.

Step 5. Separate the two wires holding the bow and slip one on each side of the carnation stem. Twist the wires twice at the stem base of the carnation to attach the bow to the stem. Cut the wires to less than an inch in length. Lay the wires against the stem and completely cover them by taping (see Figure 9-9). Shape the bow and if necessary, bend the carnation slightly forward so that the flower is facing the viewer.

Step 6. The base of the corsage may be finished in one of several ways. Cut it short and place a small hook at

FIGURE 9-8

Tape a sprig of leatherleaf fern to the back of the corsage.

FIGURE 9-9

Attach the bow to the corsage stem.

FIGURE 9-10

Corsage stem treatments.

FIGURE 9-11

The corsage may be enhanced with the addition of a sprig of baby's breath.

the bottom, curl it around a pencil, or shape into a single curve (Figure 9-10).

Step 7. Mist the flower with water, place it into a corsage bag, and close the bag with a corsage pin. Place the corsage in the refrigerator.

Step 8. Clean the work area, and evaluate the corsage using the criteria in Figure 9-6.

The boutonniere and the single-bloom carnation corsage can be modified by taping a sprig of baby's breath to the flower before the foliage and bow are added (Figure 9-11).

Constructing a Single-Spray Miniature Carnation Corsage

Step 1. Select materials (Figure 9-12):

5 mini-carnations (pompon chrysanthemums may be substituted)

5 small sprigs of leatherleaf, ming, or other foliage

5 small sprigs of baby's breath

FIGURE 9-12

Select materials for a single spray mini-carnation corsage.

FIGURE 9-13

Tape baby's breath and greenery to each carnation.

> 1-1/2 yards of number 3 or 1-1/2 satin ribbon
> 22- or 24-gauge wire
> stem wrap
> corsage bag and pin

Step 2. Wire each mini-carnation using the piercing method.

Step 3. Tape a small piece of baby's breath and leatherleaf to each of the mini-carnations (Figure 9-13).

Step 4. Select the smallest carnation as the beginning flower. Attach a second carnation to the first with a couple of twists of tape (Figure 9-14). The distance between these two flowers should be greater than the distance between other flowers. As you approach the focal point, the distance between flowers will be less.

Step 5. Add a third carnation in a staggered pattern. Secure with tape (Figure 9-15).

Step 6. Tie a bow using satin ribbon, and add the bow below the third carnation (Figure 9-16). Cut the wire to 1/2 inch in length and cover it with floral tape.

FIGURE 9-14
Tape two carnations together.

FIGURE 9-15
Add a third carnation.

FIGURE 9-16
Attach the corsage bow.

Step 7. Add a fourth carnation near the center of the bow, and tape it in place (Figure 9-17).

Step 8. Bend the fifth carnation as shown in Figure 9-18. Add this carnation among the bow loops, facing downward (Figure 9-19). Adjust the flowers and loops.

FIGURE 9-17
Add a fourth carnation near the center of the bow.

FIGURE 9-18
Bend the stem of the fifth carnation at a sharp angle.

FIGURE 9-19
Add the fifth carnation among the bow loops.

FIGURE 9-20

Bend the corsage stem into a loop.

Step 9. Wrap the stem of the corsage so that it is smooth from top to bottom. Cut the stem to about 3 inches in length and bend into a loop as shown in Figure 9-20.

Step 10. Mist the corsage with water and place in a corsage bag. Close the bag with a corsage pin. Refrigerate.

Step 11. Clean the work area, and evaluate the corsage using the criteria in Figure 9-6.

Constructing a Double-Spray Miniature Carnation Corsage

A double-spray corsage involves a larger number of flowers. This will allow you to experiment with different styles of design (Figure 9-7). The double-spray corsage may be worn in the straight-line style or shaped into a crescent or an S-curve.

FIGURE 9-21

Select materials for a double-spray corsage.

FIGURE 9-22

Tape the two sprays together into a single stem.

Step 1. Select materials (Figure 9-21):

> 9 mini-carnations (roses or pompons may be substituted)
> 9 small pieces of leatherleaf or other foliage
> 9 sprigs of baby's breath
> 1-1/2 yards of number 3 satin ribbon
> 22- or 24-gauge wires
> stem wrap or floral tape
> corsage bag and corsage pin

Step 2. Wire each of the carnations using the piercing method and add a small sprig of baby's breath and greenery as previously instructed in making a single-spray corsage (Figure 9-13).

Step 3. Make two stems of three flowers each as instructed earlier (Figure 9-15).

Step 4. Tape the two sections together, face to face (Figure 9-22).

Step 5. Bend the bottom section down sharply over your thumb (Figure 9-23).

Step 6. Add the bow in the center area between the two stems (Figure 9-24).

Step 7. Add the two remaining flowers among the loops of the bow to connect the top and bottom sections (Figure 9-25).

FIGURE 9-23
Bend the bottom spray down sharply over your thumb.

FIGURE 9-24
Add the bow.

FIGURE 9-25
Add the two remaining flowers.

FIGURE 9-26
Work the corsage into the desired shape.

FIGURE 9-27
An S-curve wrist corsage.

Step 8. Shape the corsage into the desired style: straight-line, crescent, or S-curve (Figure 9-26).

Step 9. A double-spray corsage shaped into an S-curve makes an excellent wrist corsage (Figure 9-27).

Step 10. Clean the work area, and evaluate the corsage using the criteria in Figure 9-6.

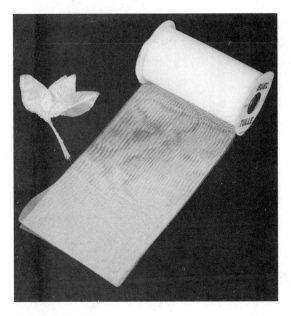

FIGURE 9-28
Netting and artificial leaves.

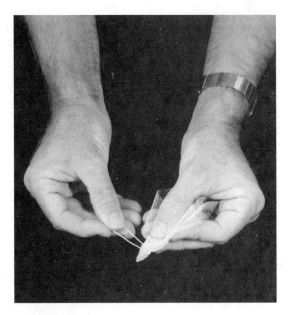

FIGURE 9-29
Making a net tuft.

NET TUFTS

In corsage construction, florists sometimes use a **net tuft**, a small piece of net, gathered in the middle with a wire and artificial leaves. Netting, also called **tulle**, comes in bolts that are 6 inches wide and 25 yards long. Netting and artificial leaves come in a variety of colors to match the color of the flowers used in the corsage or the dress of the wearer (Figure 9-28).

Making Net Tufts

Step 1. Cut a piece of tulle 6 inches in length.

Step 2. Pick up the net at the center of the square and gather it between your fingertips.

Step 3. Twist a 24-gauge wire around the net about 1/2 inch from its end (Figure 9-29). The wire must be twisted tightly or it will slip off.

Step 4. Use stem wrap to tape the wire and the tip of the net.

Constructing a Single-Carnation Corsage Using Net Tufts and Artificial Leaves

Step 1. Select materials:

> 1 standard carnation
> 3 net tufts
> 2 artificial leaves
> 1-1/2 yards of number 1-1/2 or 3 ribbon
> wire
> stem wrap

Step 2. Place the two artificial leaves onto the carnation in a diagonal pattern and tape them in place (Figure 9-30).

Step 3. Place the three net tufts around the carnation and spread the net so that the flower is surrounded by it (Figure 9-31).

Step 4. Add the bow and slightly curl the leaves. Cut the stem of the corsage to 3 inches and curl the stem. Tip the face of the carnation forward (Figure 9-32).

Step 5. Clean the work area, and evaluate the corsage using the criteria in Figure 9-6.

FIGURE 9-30
Tape artificial leaves to the carnation.

FIGURE 9-31
Add net tufts to the corsage.

FIGURE 9-32
The completed corsage.

FOOTBALL MUM CORSAGES

In some areas of the United States, football mum corsages are popular at college and high school football games. These are especially popular for homecoming games and dances. Football mum corsages may be designed simply or elaborately. In some areas, such as Texas and Oklahoma, large, extravagant designs are popular. These include an abundance of streamers, braided ribbons and accessories. A simple basic design will be demonstrated in this unit. Additional streamers and accessories may be added for a more elaborate design.

Constructing a Football Mum Corsage

Step 1. Select materials:

> 1 football mum (live or silk)
> number 3 satin ribbon in school colors
> 1 chenille stem in school color

Step 2. Tie a large bow of the darker school color using number 3 satin ribbon. The bow should contain an abundance of loops of equal length. The bow should be large enough to extend beyond the edge of the mum. Add several streamers 12 to 15 inches in length (Figure 9-33).

Step 3. Cut the stem from a football mum to about 1 to 1-1/2 inches in length. Spray the base of the mum with a spray adhesive to prevent the petals from shattering. Wire with a 20-gauge wire using the hook method. Wrap the stem and wire with floral tape. A silk mum may be substituted if desired. If the stem of the silk mum is stiff, remove it and make a stem by wrapping three 20-gauge wires together with floral tape. Add a little hot glue to the tip of the wires to secure the wire stem to the mum.

Step 4. Tie a second smaller bow of number 3 satin ribbon in the other school color. Add several streamers 12 to 15 inches long to this bow also.

Step 5. Place the mum in the center of the first bow and tie the bow to the corsage stem. Add the second bow to the front of the corsage (Figure 9-34). Adjust the

FIGURE 9-33

Tie a large bow of the darker school color.

FIGURE 9-34

Attach the bows to the mum.

FIGURE 9-35

Add the school letters to the mum.

FIGURE 9-36

Attach accessories.

bow loops and streamers around the mum. Cut all wires to about 1 inch in length and wrap them with floral tape to the mum stem. Bend the mum head slightly forward.

Step 6. Shape letters out of chenille stems in one of the school colors. With a glue gun, add a small amount of low-temperature glue to the back of the letters and place them onto the center of the mum quickly before the glue cools (Figure 9-35).

Step 7. Glue accessories to the mum such as miniature megaphones, football helmets, and footballs. These may also be tied to the streamers (Figure 9-36).

Step 8. Curl the corsage stem and stick two corsage pins into the stem. Carefully place the corsage in a corsage box so the streamers are not damaged.

Step 9. Clean the work area, and evaluate the corsage using the criteria in Figure 9-6.

DESIGNING CORSAGES USING FLORAL ADHESIVE AND HOT GLUE

An alternative method to wiring and taping corsages is the use of floral adhesives or glue. **Floral adhesive** is a rubber cement developed for use on fresh flowers. The harmful chemicals in it have been removed. Floral adhesive may be used in corsages of very light and delicate flowers, such as snapdragon blossoms. It is also used to add greenery, flowers, and accessories to a wired and taped corsage.

Corsages may also be constructed using hot glue. **Hot glue** may be purchased in several forms, including pan-melt glue chips, hot-melt glue sticks, and low-temperature glue sticks, also known as cold-melt glue sticks.

Glue sticks may be melted in an electric fry pan (Figure 9-37), glue pot, or electric glue gun. The hot glue melts at a temperature of about 275°F. Low-temperature glue sticks melt at a much lower temperature which reduces the possibility of injuries. Low-temperature glue sticks are also less likely to damage delicate flowers, but they do not have the holding strength of the hot glue.

Caution: Use caution when using hot glue. Severe burns can result from contact with your skin.

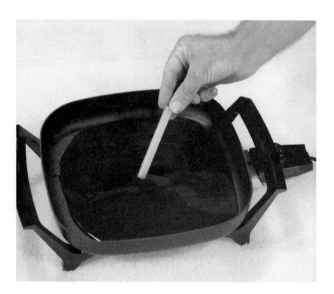

FIGURE 9-37

Melting hot glue in an electric fry pan.

Constructing a Corsage Using Low-Temperature Glue

Step 1. Select materials:

> 1 carnation
> 5 small sprigs of statice
> 1-1/2 yards of number 1-1/2 satin ribbon
> 5 small sprigs of ming or sprengeri fern
> low-temperature glue sticks
> 20- or 22-gauge wire
> stem wrap

Step 2. Tie a bow using number 1-1/2 satin ribbon. The wire used to hold the ribbon will also serve as the stem of the corsage. Use a larger wire such as 20- or 22-gauge wire so the stem will have the strength to support the corsage. All the loops in the corsage should be the same length and the bow should be rounded and about four inches in diameter. Press the center of the bow to flatten it.

Step 3. Begin the corsage by dipping the ends of small sprigs of ming or sprengeri in the hot glue and attaching them to the center of the bow (Figure 9-38). Hold the sprigs in place until the glue cools.

Step 4. Glue small sprigs of statice to the center of the corsage (Figure 9-39).

FIGURE 9-38
Glue sprigs of greenery to the center of the bow.

FIGURE 9-39
Add small sprigs of statice.

FIGURE 9-40
Glue the carnation to the
center of the corsage.

FIGURE 9-41
The completed corsage.

Step 5. Remove the stem from the carnation. Dip the base
of the carnation in glue and place it in the center
of the corsage (Figure 9-40).

Step 6. Add any additional sprigs of greenery that are needed
to fill open spaces and cover mechanics. Curve the
stem, and the corsage is complete (Figure 9-41).

This technique can be used to construct a variety of corsages. Glueing is especially effective when light, delicate
flowers are used (Figure 9-42). Wiring several small flowers

FIGURE 9-42
Glueing is effective
when light, delicate
flowers are used.

FIGURE 9-43

Netting can replace the bow in a corsage which has been constructed with hot glue.

for a corsage can be time consuming and these lightweight flowers are easily held in place by the glue.

A similar corsage can be constructed by tying a bow from net and repeating the above steps (Figure 9-43).

Student Activities

1. View any available video tapes on designing corsages.
2. Complete each of the designs illustrated in this unit.
3. Invite a florist to demonstrate basic corsage designs to the class.

Self-Evaluation

A. Fill in the blank space with the word that best completes each of the following statements.

1. Flowers worn by women are called _____.
2. Flowers worn by men are called _____.

3. The most popular flowers for boutonnieres are _____ and carnations.

4. Corsages are commonly worn on the woman's _____ shoulder.

5. Corsages may also be worn on the lady's purse, in her hair, and on her _____.

6. Corsage designs include the crescent, the straight line, triangular, round, and the _____.

7. Mums are sprayed with a spray adhesive to prevent the petals from _____.

B. Short Answer Questions

1. What criteria should be used in selecting corsage flowers, colors, and design?

2. List the supplies that would be needed in a corsage-making area.

3. How does the size of the wearer affect the choice of flowers and the style of the corsage?

Bud Vases

Terms to Know

bud vase
cardett

Materials List

*an assortment of bud
   vases*
*cut flowers (roses,
   carnations,
   gladiolus, baby's
   breath, or other
   suitable flowers)*
*cut greens (leatherleaf,
   sprengeri, plumosa,
   ming)*
floral preservative
leaf polish
ribbon
20- or 22-gauge wire

OBJECTIVE

To design a bud vase.

Competencies to Be Developed

After completing this unit, you should be able to:

- design a one-bloom bud vase.
- design a three-bloom bud vase.
- design a bud vase for special occasions.

Introduction

A **bud vase** is a tall, narrow container with a small neck designed to hold one or a small number of flowers. They are quick and easy to make and are modestly priced. Because of this, bud vase arrangements are popular with customers. They make excellent gifts for hospital patients and are very popular on holidays, such as Mother's Day and Valentine's Day.

Bud vases are made of various materials and come in a variety of shapes (Figure 10-1). Some of the most popular materials are clear glass, milk glass, colored glass, silver, plastic, pottery, and china. The cheaper glass and plastic vases are the most popular in the flower shop. The more expensive

FIGURE 10-1
A variety of bud vases.

materials greatly increase the cost of the bud vase arrangement. Customers who are willing to pay the increased cost usually choose a different type of arrangement. Expensive bud vases are often purchased by customers and kept in the home to be used over and over.

The color of the bud vase chosen should harmonize with the flower colors. Clear, milk, and green glass, as well as white and green plastic, are popular because they can be used for most flowers and occasions.

The neck of bud vases may vary in size to accommodate varying numbers of flowers. A bud vase for a single blossom should have a small neck to support the stem and prevent the flower from moving around in the vase. Bud vases with a narrow neck would not accommodate the bulk of multiple flowers. Choose a neck designed for the number of flowers you plan to use.

SELECTING FLOWERS FOR BUD VASES

Roses and carnations with baby's breath are the most popular flowers used in bud vases. However, any number of flowers are suitable. The main requirement is that the flower have a stem long enough to be inserted into the vase and still hold the flower well above the rim of the container.

In the spring, iris and daffodils are in season. At other times of the year, standard mums make excellent novelty bud vases (Figure 10-2). Gladioli, placed back to back, create a contemporary look (Figure 10-3). Because of the weight of these flower heads, the bud vase will need to be weighted with marbles or gravel. A large neck vase will also be required.

FOLIAGE FOR BUD VASES

Foliage complements the flowers in bud vases and helps to hide stems and mechanics if any are used. Foliage also helps hold flowers in place.

Leatherleaf is a popular foliage used by florists when the bud vase is to be viewed from one side only. A bud vase to be taken to a hospital would probably only be viewed from one

FIGURE 10-2
Mum novelty bud vase.

FIGURE 10-3
Gladiolus bud vase.

FIGURE 10-4
Bud vase with
eucalyptus foliage.

side. When the bud vase is to be viewed from any direction, other foliages, such as ming, plumosa, or sprengeri are more appropriate. Foliage such as eucalyptus, huckleberry, and Scotch broom contribute to the distinctive design of the bud vase (Figure 10-4).

ACCESSORIES

Bows and other accessories are often used on the bud vase. If the bud vase is made for a special occasion, the accessories help contribute to the theme. For example, if the occasion were a Fourth of July celebration, then a red, white, and blue ribbon would complement the theme. A small flag would be an additional thematic accessory.

Shamrocks on St. Patrick's Day would be an ideal accessory. Butterflies and bumblebees are often used on novelty bud vases.

DESIGNING BUD VASES

Constructing a One-Bloom Rose Bud Vase

Step 1.   Select materials:

> 1 long-stemmed rose of any color
> 3 stems of sprengeri or ming foliage
> 1-1/2 yards of number 3 satin ribbon
> 2 20- or 22-gauge wires
> 1 stem of baby's breath (optional)
> leaf polish
> bud vase

Step 2.   Fill a bud vase with preservative water to within 2 inches of the top.

Step 3.   Remove all foliage that will be below the water level. To take off thorns, a rose stripper is useful (Figure 10-5).

Step 4.   Cover the rose bloom with your hand, and spray the foliage with leaf polish to make it shine.

Step 5.   Cut the flower stem below water to a length 1-1/2 to 2 times the height of the vase.

Step 6.   Wire the rose with a 20- or 22-gauge wire using the straight-wire method. Gradually wind the wire around the stem of the rose, being careful not to break off leaves.

Step 7.   Insert the flower in the vase and turn the "face," or most desirable side, toward you.

FIGURE 10-5

Use a rose stripper to remove thorns and foliage.

Step 8. Cut three pieces of sprengeri or ming foliage to different lengths and place these so that they arch slightly away from the rose (Figure 10-6).

Step 9. Add a sprig of baby's breath so that it surrounds the rose inside the foliage.

Step 10. Tie a bow about the size of a corsage bow. Leave three streamers cut to different lengths up to 6 inches long. Tie the bow using a 20-gauge wire. If smaller wire is used, attach it to a wooden pick.

Step 11. Insert the bow in the front center of the bud vase slightly above the rim. Adjust the streamers.

Step 12. Attach a card to the ribbon or insert a **cardett** and card (Figure 10-7). A cardett is a plastic stem with a three-prong holder at the top to hold a card.

Step 13. Refrigerate the completed bud vase.

FIGURE 10-6
Add sprengeri or ming fern to the bud vase.

FIGURE 10-7
Attach a cardett to the bud vase.

Constructing a Three-Bloom Carnation Bud Vase

Step 1. Select materials:

3 carnations
5 stems of leatherleaf fern
1 large sprig of baby's breath
floral preservative
leaf polish
1-1/2 yards of number 3 satin ribbon
20- or 22-gauge wire
a large-necked bud vase

Step 2. Prepare floral preservative, and fill the bud vase to within 2 inches of the top.

Step 3. Spray leatherleaf fern with leaf polish.

Step 4. Wire the carnations using the straight-wire method. Flowers may not need wiring if the stems are strong.

Step 5. Select the smallest carnation and cut it to twice the height of the container.

Step 6. Select the medium-sized bloom and cut it one to two inches shorter than the tallest flower.

Step 7. Cut the largest flower one to two inches shorter than the middle flower.

Step 8. Remove all foliage from the stems that will be below the vase line.

Step 9. Insert the flowers in the bud vase. Place the smallest flower in the center, facing up. Stagger the medium flower and place the largest flower at the bottom center (Figure 10-8). Gradually face the middle and bottom flowers toward the viewer.

Step 10. Select two pieces of leatherleaf which are slightly taller than the flowers, and place these back-to-back behind the flowers. This causes the leatherleaf to stand up straight behind the flowers.

Step 11. Insert shorter pieces of leatherleaf on each side at a 45° angle and in front of all other stems. An op-

FIGURE 10-8
Placement of the carnations in the bud vase.

FIGURE 10-9

Add leatherleaf to the bud vase.

FIGURE 10-10

Add baby's breath and attach a bow.

tional piece may be inserted vertically in front of the tallest stem (Figure 10-9).

Step 12. Insert baby's breath and add a bow centrally, slightly above the front rim of the vase (Figure 10-10). The bow is attached to a wooden pick that is inserted into the bud vase.

Step 13. Add a card and refrigerate.

Student Activities

1. Make the one-bloom and three-bloom bud vases as outlined in this unit.

2. Design a bud vase for a special occasion. Be creative and use any accessories you feel would add to the theme of the occasion.

3. Invite a florist to visit your class and demonstrate the creation of unique bud vases.

4. Use the Flower Arrangement Rating Scale in Appendix G to rate one of the bud vases that you designed.

Self-Evaluation

A. Fill in the blanks with the word that best completes each of the following statements.

1. A _____ is a tall narrow container with a small neck.

2. Bud vases are popular with customers because they are less _____ than other types of arrangements.

3. The color of the bud vase chosen should _____ with the colors of the flowers to be used in the bud vase.

4. The most popular flowers used in bud vases are _____ and _____.

5. Leatherleaf is a popular foliage used in designing bud vases, but it should be used only when the arrangement is to be viewed from _____.

6. Bows, shamrocks, and butterflies are examples of _____ that may be used on bud vases.

7. The overall height of the bud vase arrangement should be _____ the height of the bud vase.

8. The purpose of a cardett is to attach a _____ to the bud vase arrangement.

B. Short Answer Questions

1. Why do retail flower shops mostly use inexpensive plastic and glass bud vases?

2. Why are the necks of bud vases in different sizes?

3. How do accessories contribute to the theme of a bud vase?

Circular Arrangements

OBJECTIVE

To design circular arrangements.

Competencies to Be Developed

After completing this unit, you should be able to:

- identify four types of circular mass arrangements.
- construct a circular mound arrangement.
- construct a conical arrangement.
- construct an oval arrangement.
- construct a fan arrangement.

Terms to Know

conical design
fan arrangement
mound design
nosegay arrangement
oval arrangement

Introduction

The mass designs are one of the more popular styles of arrangements seen in flower shops today. Mass arrangements consist of many flowers arranged in a geometric pattern.

A beginning designer should visualize the geometric pattern of a design and how the completed design will look before beginning an arrangement. These patterns become a framework that guides the designer in completing the arrangement. Creativity can be expressed within this framework by varying the size of the arrangement, the compactness of the design, and the amount of depth in the arrangement.

Sometimes we think of these designs as being modern, but they actually had their origins many centuries ago. Arrangements such as the cone or conical design, developed during the Byzantine period (A.D. 320–600), are still popular designs today. The German Biedermeier design is similar in shape and consists of compact, concentric layers of different flowers and foliages (Figure 11-1). The massive designs, developed during the Victorian period in England during the 1800s and the Williamsburg period in colonial America, influenced the development of mass designs popular today.

Mass designs are based on either the triangle or the circle. This unit deals with circular designs while Unit 12 concentrates on triangular ones.

FIGURE 11-1

The Dutch Biedermeier design.

Cone

Mound

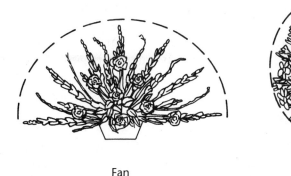

Fan

Oval

FIGURE 11-2

Design patterns for circular designs.

Four variations of the circle are presented in this unit—mounds, cones, ovals, and fans. These circular design forms are presented in Figure 11-2.

THE CIRCULAR MOUND DESIGN

The **mound design**, also called a **nosegay arrangement**, is designed to be viewed from all sides. For this reason, the mound does not have a focal point. The design may be compact (typical of the Biedermeier or Victorian style) or light and airy (typical of the French period).

Radiation rhythm is important in the construction of a mound. All of the stems are the same distance from the center of the design. They should appear to radiate from one point with the mound forming a sphere (Figure 11-3). A line drawn across the top of the design forms a half circle.

Because the mound does not have a focal point, size rhythm (placing small flowers farther from the focal point

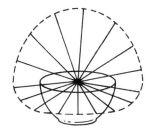

FIGURE 11-3

Stems should radiate from one point in a mound design.

and large flowers near the focal area) is not used in this design. Space rhythm (placing flowers closer together as you near the focal point) is not used either. In this design, the space between flowers is equal throughout the design.

Mound designs are usually arranged in low, round containers. If the container is taller than it is wide, an oval design would be more appropriate for the container.

Flowers used in the mound design should be evenly distributed over the arrangement. Only mass, filler, and sometimes form flowers should be used in constructing the mound. Line flowers should never be used or the design will have a spiked appearance.

Select one type of mass flower to serve as the primary flower, and place all of them into the design at one time. Add the secondary or filler flowers afterwards.

Mass and filler flowers such as carnations, roses, tulips, mums, statice, and baby's breath are recommended for mound designs. The size of the flowers will be dictated by the size of the mound. Avoid using large flowers to make a small mound design because of the difficulty of spacing large flowers in a small arrangement while maintaining the circular design. There are many ways of designing the same arrangement. The following steps represent one approach. After mastering this approach, be creative and try a different method.

Constructing a Circular Mound Arrangement

Step 1. Select materials:

> low, round container
> 1/3 block of floral foam
> anchor tape
> 9 carnations
> 3 stems of pompons
> 3 stems of baby's breath
> 8 stems of leatherleaf fern
> preservative

Step 2. Prepare the container. Select a low, flat container. Cut approximately 1/3 of a block of floral foam. If a knife is not available, use an 18-gauge wire and cut the floral foam by holding the wire in both

FIGURE 11-4

Secure the floral foam to the container.

FIGURE 11-5

"Green" the base of the arrangement.

hands and slice down through the block. The floral foam should extend about 1 inch above the rim of the container. Saturate the foam in a warm preservative solution. Secure the floral foam to the container with anchor tape. Crisscross the tape over the foam and attach the ends to the container (Figure 11-4). Do not attach more than an inch of the tape to the container. Long strips of tape will be difficult to hide later.

Step 3. "Green" the base of the arrangement by placing a circle of foliage, such as leatherleaf fern, around the rim of the container. Allow the greenery to overlap the side of the container (Figure 11-5). Use the tips of the leatherleaf fern. Save the remaining part of the stem for later use. Many designers green the entire arrangement prior to the placement of any flowers. In this demonstration, we will place greenery in the arrangement as we complete the design. Experiment and decide which suits you best.

Step 4. Establish the height of the design. Place the first carnation, the primary flower, in the center of the floral foam at a height of 1-1/2 to 2 times the width of the container (Figure 11-6).

FIGURE 11-6

Establish the height of the arrangement.

FIGURE 11-7

Add four carnations at the rim of the container.

Step 5. Add the base flowers. Place four carnations straight into the foam, opposite each other at the rim of the container. These four carnations should form a square pattern (Figure 11-7). Remember, these carnations and the first carnation should be equidistant from the center of the arrangement.

Step 6. Add additional greenery, placing one piece on each side of the center carnation. Also add the base stems from which you cut the tips in step 3 (Figure 11-8).

Step 7. Add remaining primary flowers. Insert four additional carnations at 45° angles halfway between the two base carnations and the center carnation (Figure 11-9). Vary the height of these carnations a little to prevent creating a row effect. These flowers should also be equidistant from the center of the arrangement.

Step 8. Add secondary flowers. Place pompons in the space between the carnations. Use large blossoms in large spaces and small blossoms or buds in small spaces. Do not crowd the arrangement. Leave some space around each flower (Figure 11-10).

FIGURE 11-8
Add additional greenery.

FIGURE 11-9
Add the remaining primary flowers.

Step 9. Add filler flowers. This arrangement could be used as it is or filler flowers such as baby's breath or statice can be added to soften the design. Cut small pieces of baby's breath and carefully place them

FIGURE 11-10
Add secondary flowers.

FIGURE 11-11
Baby's breath may be added to the arrangement.

between the flowers in the arrangement. Allow the baby's breath to stick out slightly around the flowers and foliage (Figure 11-11).

Step 10. Check your work. Step back and examine the arrangement from all angles. Check for holes and add greenery or flowers where they are needed. Move flowers, if necessary, to create equal space between flowers. If you move a flower, insert it into a new spot in the floral foam, making a new hole. This prevents the stem from being placed in an air pocket that will not allow water absorption.

Step 11. Evaluate the design. Use the rating scale illustrated in Appendix G to evaluate your arrangement.

Many variations of the circular mound design are possible. Vary the flowers and foliage, and arrange them creatively. Figure 11-12 shows a similar design using two colors of daisy pompons. In this arrangement the darker-colored pompon serves as the primary flower and the lighter pompon as the secondary.

FIGURE 11-12

A mound design of pompons.

THE CONICAL DESIGN

The **conical design** was first introduced during the Byzantine period (A.D. 320–600). Foliage was arranged in containers to resemble cone-shaped trees. Flowers and fruit were often spiraled from the base of the arrangement to the top.

This conical design is also popular today. The design is often varied from its original design to appear less formal. The cone is similar in shape to the pyramids even though it is round. Cut six triangles out of poster board and glue them together. The resulting structure illustrates the basic shape of the conical design.

Cone-shaped topiaries of boxwood, often used at Christmas, illustrate this design. The design can also be used to create table arrangements (Figure 11-13).

FIGURE 11-13

A conical table arrangement.

A low, round container is used for this design when the arrangement is to be used as a table centerpiece. Table centerpieces over 14 inches in height restrict visual contact across the table. Pedestaled containers give this arrangement a more dramatic look.

Constructing a Conical Arrangement

Step 1.   Select materials:

> 11 carnations
> 2 stems of pompons
> 8 stems of leatherleaf
> 3 or 4 stems of baby's breath
> a pedestal container
> 1/3 block of floral foam
> anchor tape
> preservative

Step 2.   Prepare the container. Cut approximately 1/3 of a block of floral foam. The foam should extend about an inch above the rim of the container. Saturate the foam in a warm preservative solution, then secure it to the container with anchor tape. Crisscross tape over the foam and attach the ends to the container (Figure 11-14). Do not attach more

FIGURE 11-14

Secure the floral foam to the container.

FIGURE 11-15
Green the base of the arrangement.

than 1 inch of the tape to the container. Long strips of tape will be difficult to hide later.

Step 3. Green the base of the arrangement. Place a ring of leatherleaf in a circle at the rim of the container (Figure 11-15). Use the tips and save the base of the stems for later use.

Step 4. Establish the base. The primary flower for this arrangement is carnations. Place five carnations equidistant from one another at the rim of the container (Figure 11-16). Do not be concerned about having to reposition flowers. However, do not use the same hole in the floral foam. Make a new hole each time you move a flower.

Step 5. Establish the height. Place a carnation in the center of the floral foam at a point about two times the height of the container (Figure 11-17). Designers often increase the height of the design for a more dramatic appearance.

Step 6. Add the remaining primary flowers. Add five additional carnations between the top flower and the base flowers. Vary the height of these flowers and do not extend them beyond an imaginary line from

FIGURE 11-16

Place five carnations equidistant from one another at the rim of the container.

FIGURE 11-17

Establish the height of the design.

FIGURE 11-18

Add the remaining primary flowers.

FIGURE 11-19

Add additional greenery.

the base flowers to the top flower. This will help you maintain the conical shape (Figure 11-18).

Step 7. Add additional greenery. Place two stems of leatherleaf back to back on each side of the center stem. Add additional greenery until the mechanics are covered (Figure 11-19). Remember to use the bottom of the stems used earlier to green the base of the design.

Step 8. Add secondary flowers. Insert blossoms of daisy pompons where there are large spaces in the arrangement (Figure 11-20). Do not crowd the flowers.

Step 9. Add filler flowers. This arrangement could be used as it is, or filler flowers such as baby's breath or statice can be added to soften the design. Cut small

FIGURE 11-20
Add secondary flowers.

FIGURE 11-21
Baby's breath can be added to the arrangement.

pieces of baby's breath and carefully place it between the flowers in the arrangement. Allow the baby's breath to stick out slightly around the flowers and foliage (Figure 11-21).

Step 10. Check your work. Step back from the arrangement and observe it from all sides. Move flowers if you wish, but remember to make a new hole in the floral foam each time you do. When the design is completed to your satisfaction, lightly mist the flowers and foliage to give them a fresher appearance.

Step 11. Evaluate the design using the rating scale in Appendix G.

THE OVAL ARRANGEMENT

The **oval arrangement** tends to hold the viewer's eye within the circular pattern created by the design. For this reason, it is sometimes used at the altar of a church or at the end of a room. Large oval designs are also used at wedding receptions and parties.

Constructing an Oval Arrangement

Step 1. Select materials:

a tall vase or pedestaled container
1/2 block of floral foam
anchor tape
13 carnations
4 stems of miniature carnations
4 stems of pompon chrysanthemums
1 bunch of leatherleaf
half of a bunch of Boston fern

Step 2. Prepare the container. Cut 1/3 to 1/2 of a block of floral foam. Saturate in warm preservative solution. Trim the bottom half of the block so that it fits into a pedestaled container, and secure it with anchor tape (Figure 11-22).

Step 3. Green the base of the arrangement. Place a circle of leatherleaf around the rim of the container, allowing the greenery to drape down over the side of the container (Figure 11-23).

Step 4. Establish the height of the design. Place the first carnation in the center of the floral foam at a height of 1-1/2 to 2 times the height of the container (Figure 11-24).

Step 5. Add the base flowers. Place four carnations equidistant from each other at the rim of the container. Allow these carnations to drape over the edge of the container (Figure 11-25).

Step 6. Add remaining primary flowers. The remaining primary flowers will be added in two stages. Select four carnations with arching stems, if possible. Place these at equal distances from each other below the

FIGURE 11-22

Trim and secure the floral foam to the container.

FIGURE 11-23
Green the base of
the arrangement.

FIGURE 11-24
Establish the height of the design.

FIGURE 11-25
Add the base flowers.

center carnation. Stems should not cross but radiate from the center of the floral foam. Vary the height of each stem slightly so that the flowers do not create rows (Figure 11-26). Place four additional carnations halfway between the top carnation and the base carnations. Again, select carnations with arching stems if possible. Stagger these flowers between the base carnations (Figure 11-27).

Step 7.   Add secondary flowers. Add miniature carnations in the spaces between the standard carnations. There are several flowers per stem, so remove all but one or two of the top flowers on the main stem. Use these high in the arrangement, and,

FIGURE 11-26
Add four carnations below the center carnation.

FIGURE 11-27
Add four additional carnations to the arrangement.

lower in the design, use the shorter-stemmed flowers that were removed earlier (Figure 11-28).

Step 8. Add additional greenery. Greenery with graceful, arching stems, such as sprengeri fern, is excellent for an oval arrangement. In this demonstration, we will be adding some additional leatherleaf and stems of Boston fern. Fill in the spaces between flowers, but do not crowd the arrangement. The foliage should appear to radiate from the center (Figure 11-29).

Step 9. Add filler flowers to the arrangement. Baby's breath is used here but statice, spray asters, or other filler flowers may be substituted (Figure 11-30).

Step 10. Check your work. Step back from the arrangement, and observe it from all sides. Move flowers as

FIGURE 11-28
Add secondary flowers.

FIGURE 11-29
Green the arrangement.

FIGURE 11-30

Add filler flowers such as baby's breath.

needed, but remember to make a new hole in the floral foam each time.

Step 11. Evaluate the design, using the rating scale in Appendix G.

THE FAN ARRANGEMENT

The **fan arrangement** is the first one-sided arrangement that we will design. It obtains its name from the open fan silhouette created by the design.

Fan arrangements, typical of many Williamsburg designs, are still popular. They are often used as flowers for the church altar, banquet arrangements, and basket designs for sympathy flowers.

Constructing a Fan Arrangement

Step 1. Select materials:

> 1 bunch of snapdragons
> 12 carnations
> 3 stems of pompons
> 3 stems of statice
> 1 bunch of leatherleaf
> 1/2 block of floral foam

Step 2. Prepare the container. Cut 1/2 of a block of floral foam. Saturate the foam in a warm preservative solution. Secure it to the container with anchor tape.

Step 3. Use seven snapdragons to create the fan silhouette as shown in Figure 11-31. The two lower snaps should angle over the rim of the container.

FIGURE 11-31

Use seven snapdragons to create the fan silhouette.

Step 4. Place the three remaining snapdragons inside the silhouette near the focal area (Figure 11-32). The placement of linear flowers in this manner will help to integrate the outer line flowers with the mass flowers.

Step 5. Place a large carnation at the lower center of the design so that the flower extends out from the container several inches. This will help create depth in the arrangement. Place five to seven additional carnations at the outer edges of the arrangement. Wire any carnations with weak stems. Place additional carnations in the area between the center carnation and the outer carnation (Figure 11-33).

Step 6. Add greenery to the design, following the fan shape of the arrangement. In this illustration, we have used leatherleaf, with a small amount of Scotch broom and eucalyptus to add interest to the design (Figure 11-34). Also, filler flowers could be added if you wish to do so.

FIGURE 11-32

Add three snapdragons inside the silhouette near the focal area.

FIGURE 11-33
Add carnations to the arrangement.

FIGURE 11-34
Green the arrangement.

Step 7. Add greenery to the back of the arrangement. Even though this is a one-sided design, the back of it is often visible. All mechanics should be covered.

Step 8. Check your work. Step back from the arrangement and observe it closely. Move flowers as needed, but remember to make a new hole in the floral foam each time.

Step 9. Evaluate the arrangement, using the rating scale in Appendix G.

Student Activities

1. Complete each of the four arrangements described in this unit according to the pattern given. After you have practiced, create your own designs without following the formula outlined.

2. Examine florist trade magazines to find examples of circular arrangements, and try copying some.

3. As a class activity, design floral arrangements to sell for special holidays occurring during the semester.

4. Plan a small flower show in your class, using circular designs. Offer modest prizes to the winner.

5. Invite a florist to your class to illustrate the designing of one or more circular designs.

Self-Evaluation

A. Flower Arrangement Rating Scale— Complete an evaluation of the arrangements using the rating scale in Appendix G.

B. Complete each of the following statements.

1. The cone arrangement was developed during the _____ period.

2. Arrangements consisting of many flowers in geometric form are called _____ arrangements.

3. Four circular design forms are the mound, cone, oval, and _____.

4. The mound design does not utilize _____ rhythm because all of the stems are the same distance from the center of the design.

5. Table centerpieces should not be over _____ inches in height.

6. Large oval designs are frequently used at _____ and parties.

7. The _____ arrangement is a one-sided arrangement.

C. Short Answer Questions

1. The mound design does not have a focal point. Why not?

2. Why are line flowers not used in a mound arrangement?

3. Which of the four circular designs would be most appropriate for a large altar arrangement?

Triangular Arrangements

OBJECTIVE

To design triangular arrangements.

Competencies to Be Developed

After completing this unit, you should be able to:

- construct an equilateral triangle arrangement.
- construct an isosceles triangle arrangement.
- construct a centerpiece arrangement.
- construct an asymmetrical triangle arrangement.
- construct a scalene triangle arrangement.
- construct a right triangle arrangement.

Terms to Know

asymmetrical triangle
centerpiece design
equilateral triangle
isosceles triangle
right triangle
scalene triangle

Introduction

Triangular arrangements are among the most basic floral designs used today. These designs are very versatile. They may be used in formal, informal, or contemporary settings. It is a design style suitable for any occasion because of its many variations.

Triangular designs, based on the classic triangular form established by the Greeks' mathematical approach to architecture, may be symmetrical or asymmetrical in form. Symmetrical triangles occur when there is equal distribution of

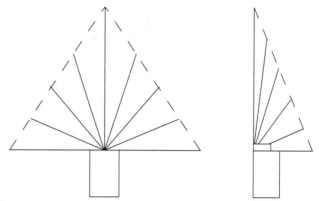

FIGURE 12-1

The triangular pattern and radiating line movement for the equilateral triangle.

materials on both sides of an imaginary center axis. Asymmetrical triangles occur when there is equal visual weight on both sides of a center axis, but the two sides do not look alike.

Triangular designs have one focal area from which all stems appear to radiate. Advancing colors should be used at the focal area and flowers should be spaced closer as they near that point. Stems should never cross. The focal point almost always rests on the rim of the container.

THE EQUILATERAL TRIANGLE

An **equilateral triangle** requires that all three sides of the design be equal in length. The triangular shape and radiating line movement for the equilateral triangle can be seen in Figure 12-1.

Constructing an Equilateral Triangle Arrangement

Step 1.   Select materials:

> a low container
> 5 gladioli
> 13 carnations
> 2 stems of statice
> camellia, lemon leaf, or huckleberry foliage
> stems of leatherleaf
> 1/2 block of floral foam

Step 2.   Prepare the container. Select a low or pedestaled container for this design. Soak floral foam in warm preservative solution and place it in the container so that the foam protrudes at least an inch above the container. Secure with anchor tape.

Step 3. Use three gladioli to establish the height and width of the arrangement. Select three more that are the least open. These will form the points of the triangle. The first gladiolus should be placed in the rear center of the floral foam and tipped slightly backwards to counterbalance the forward placement of other flowers (Figure 12-2). The height of the arrangement should be at least two times the height of the container. The other two gladiolus will be used to establish the width of the arrangement. They should be distanced so that each of the three flowers is equidistant, creating three equal triangle sides. Insert the stems into the sides of the foam near the back. The stems should angle slightly backward and downward (Figure 12-3).

FIGURE 12-2

Placement of the first line flower.

FIGURE 12-3

Placement of two additional gladioli complete the triangular pattern.

FIGURE 12-4

Add two additional gladioli at 45° angles.

FIGURE 12-5

Add carnations following the triangular pattern of the design.

Step 4. Position an additional gladiolus at the midpoints of the two vertical sides of the triangle (Figure 12-4). If you were constructing a large arrangement, more than one flower would be needed on each side. Use your own judgment.

Step 5. Add carnations following the triangular pattern of the design (Figure 12-5). Place the largest carnation in the front center, just above the front rim of the container. Position it to come straight out and slightly downward. This will create depth in the arrangement. Angle the lower flowers downward over the rim. All of the stems should appear to radiate from one point in the arrangement. If the flowers are of different sizes, place the smaller blossoms at the edges of the design and the larger ones near the focal area. Correct facing of the flowers is important in this design. Flowers near the focal area should face forward to make them appear larger.

Step 6. Green the arrangement. Insert greenery into the foam along the triangular lines of the design in

FIGURE 12-6

Gren the arrangement.

FIGURE 12-7

Add statice to complete the arrangement.

front of and behind the flowers (Figure 12-6). Green the back of the arrangement so that all mechanics are covered.

Step 7. Add statice or other filler flowers to fill in voids. Be careful not to crowd the flowers (Figure 12-7).

Step 8. Evaluate the arrangement using the rating scale in Appendix G. Multiply the total points by two to convert to a number grade.

Many variations of the equilateral triangle are possible. Vary the flowers and foliage. The use of line flowers is not necessary. The points of the triangle can be established with mass flowers.

THE ISOSCELES TRIANGLE

The **isosceles triangle** is a variation of the equilateral triangle. In this design, two sides are equal in length while the third side is shorter. The height of this arrangement is greater than the equilateral triangle, while the width is smaller. Figure 12-8 shows the design form and radiating lines of the isosceles triangle.

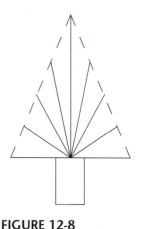

FIGURE 12-8

The triangular pattern and radiating line movement for the isosceles triangle.

**Constructing an Isosceles
Triangle Arrangement**

Step 1. Select materials:

12 carnations
1 stem of pompons
1 stem of statice
leatherleaf stems
1/3 block of floral foam

Step 2. Prepare the container. Place soaked floral foam in the container so that about an inch sticks up above the container. Secure it with anchor tape crossed over the foam.

Step 3. Establish the points of the triangle. Select three small carnations to form the points of the isosceles triangle. Place one carnation at the center back of the foam leaning slightly backward. This flower should be 2-1/2 to 3 times the height of the container. Place the other two carnations into the side of the foam near the back. Be certain that the width of the arrangement is less than the height so that an isosceles triangle is formed (Figure 12-9).

Step 4. Add additional carnations according to the pattern shown in Figure 12-10. Begin by placing the largest carnation at the lower center of the design, facing forward and slightly downward over the rim of the container. This will give depth to the arrangement and create a focal point.

Step 5. Place pompon chrysanthemums in the spaces between the carnations, and green the arrangement (Figure 12-11).

Step 6. Add statice to soften the arrangement, and create greater depth by placing some sprigs below the line of the carnations. Place the statice evenly throughout the arrangement.

Step 7. Evaluate the arrangement using the rating scale in Appendix G.

FIGURE 12-9

Use three carnations to establish the points of the triangle.

FIGURE 12-10

Add remaining carnations.

FIGURE 12-11

Add pompons in the spaces between the carnations.

CENTERPIECE DESIGNS

The **centerpiece design** is a low, horizontal design that makes an excellent table centerpiece. This arrangement can be viewed from any direction without blocking the view of a seated dinner guest. The long and narrow shape of the design can be altered to match the width of the dining table.

The centerpiece design is constructed in the same manner as a triangular arrangement but is finished on both sides. Figure 12-12 shows the triangular pattern and the radiating line movement of the design.

Constructing a Centerpiece Arrangement

Step 1. Select materials:

        container
        floral foam
        9 carnations
        3 to 4 stems of pompons
        2 to 3 stems of statice
        2 to 3 stems of baby's breath
        leatherleaf stems

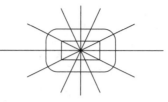

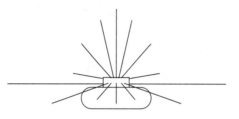

FIGURE 12-12

Radiating line movement of the centerpiece arrangement.

Step 2. Prepare the container. Since this design has a low profile, select a low oval or rectangular container. The container should be in proportion to the size of the table. Soak floral foam and place it in the container so that at least an inch sticks up above the container.

Step 3. Select five carnations to form the skeleton of the design as shown in Figure 12-13. The height of the arrangement should be no greater than 14 inches so as not to obstruct any diner's view. The length of the arrangement should be 1-1/2 to 2 times the length of the container.

Step 4. Add the remaining 4 carnations as shown in Figure 12-14.

Step 5. Green the arrangement. Begin by placing a stem of leatherleaf on both sides of the center carnations. Add mass and filler flowers to complete the arrangement (Figure 12-15 and color insert). Do not crowd the flowers.

Step 6. Evaluate the design using the rating scale in Appendix G.

FIGURE 12-13

Place five carnations to form the skeleton of the design.

FIGURE 12-14

Add remaining carnations.

FIGURE 12-15

Green the arrangement, and add mass and filler flowers to complete the design.

THE ASYMMETRICAL TRIANGLE

The **asymmetrical triangle** arrangements are more relaxed and visually pleasing designs. Asymmetrical compositions are visually balanced but differ on each side of a vertical axis. In the asymmetrical triangle, the vertical axis is not centered but usually offset to the left of center. Figure 12-16 gives the design form and radiating line movement for the asymmetrical triangle.

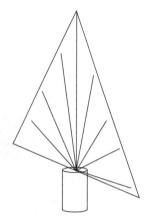

FIGURE 12-16

The triangular pattern and radiating lines for the asymmetrical triangle.

Constructing an Asymmetrical Triangle Arrangement

Step 1. Select materials:

a pedestal, or cylinder container
5 gladioli or other line flower
4 to 5 stems of pixie carnations
camellia or lemon leaf foliage
floral foam

Step 2. Prepare the container. Trim presoaked floral foam to fit the selected container. Secure with anchor tape. The floral foam should stick up at least an inch above the container.

Step 3. Place the skeleton flowers. Flower number one will establish the height of the design and should be offset to the left and rear of the foam. The asymmetrical triangle design should be two or more times the height of the container. Flower number two should be approximately two-thirds the length of the first. Place this flower to create an L-form with the first flower. Usually designers tip it slightly forward in a diagonal direction to the right. Cut the third flower to two-thirds the length of the second. Place this flower at the left edge of the foam at a 45° angle. These flowers complete the skeleton form of the design (Figure 12-17). If a line connected their tips, the asymmetrical triangle would be seen.

FIGURE 12-17

Use three gladioli to establish the triangular pattern for the asymmetrical triangle.

FIGURE 12-18

Add flowers four and five.

Step 4. Flowers number four and five may be line or mass flowers. Number four should be the tallest and placed on the high side of the skeleton. This gives visual balance to the design. Number five should be shorter and to the right of the center (Figure 12-18). The secondary flowers should never extend beyond the imaginary lines of the asymmetrical triangle created by connecting the outer points of the arrangement.

Step 5. Add mass and filler flowers. Cut individual blossoms of gladiolus on short stems and add these in the focal area with other mass or filler flowers. In this arrangement, we have chosen pixie carnations. Follow the triangular form of the design in their placement (Figure 12-19).

FIGURE 12-19

Add short stems of gladiolus and mass flowers to the design.

FIGURE 12-20

The completed arrangement.

Step 6. Green the arrangement using lemon leaf, camellia, or similar foliage to complete the arrangement (Figure 12-20). Remember to green the back of the arrangement.

Step 7. Evaluate the design using the rating scale in Appendix G.

The asymmetrical triangle allows for greater creativity. The skeleton flowers can be varied to create dramatic designs. After you have mastered this design, lengthen the number one flower and shorten the number two and three flowers to create a tall, slender design.

THE SCALENE TRIANGLE

A variation of the asymmetrical triangle is the **scalene triangle**. The scalene triangle is also composed of three unequal sides. The difference is that the number one flower is slanted to the left rather than vertical. Figure 12-21 gives the design form and radiating line movement for the scalene triangle. The number one flower establishes the height of the design. This flower is placed to the left of center on a diagonal and slants slightly backward. The number two flower is on the left and may be placed at an angle or horizontal with the table. This flower is often placed at a forward angle also. The number three flower may be angled or placed on the horizontal, depending upon the height of the container used.

Figure 12-22 and the color insert illustrate the scalene triangle. Because of the similarity to the previously designed asymmetrical triangle, a step-by-step procedure will not be presented.

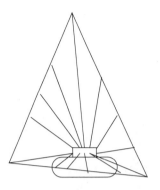

FIGURE 12-21

The triangular pattern and radiating line movement of the scalene triangle.

FIGURE 12-22

The scalene triangle.

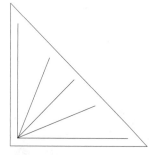

FIGURE 12-23

The triangular pattern and radiating line movement of the right triangle arrangement.

THE RIGHT TRIANGLE

The **right triangle** arrangement is a one-sided mass design that resembles half of an equilateral or isosceles triangle. Figure 12-23 shows the basic design form and radiating line movement for the right triangle. Note that the vertical and horizontal lines of the arrangement meet at a 90° angle.

The right triangle may face either the right or left. Frequently, this design occurs in pairs on a mantel or buffet. A single arrangement may be used if it is offset and visually balanced by some other object, such as a painting.

Constructing a Right Triangle Arrangement

Step 1. Select material:

a low container
floral foam
2 gladioli
4 or 5 stems of pompons
leatherleaf fern stems

Step 2. Place the floral foam in a low container and secure with anchor tape.

Step 3. Place the three skeleton flowers according to the pattern shown in Figure 12-23. The number one flower establishes the height of the design. This should be about two times the length of the container. Place this flower in the back left corner of the foam. It should point directly upward. The number two flower should be about two-thirds the length of the number one flower. Place it toward the back, in the right side of the foam. This flower may be horizontal with the table or angled slightly downward. The third flower is placed on the left front side of the foam and angled slightly forward. Since this flower is short, the right angle pattern of the design is not destroyed. Figure 12-24 illustrates the completed skeleton of the design.

Step 4. Add mass flowers. Begin by placing pompon chrysanthemums along the edge of the arrangement.

FIGURE 12-24

The skeleton for the right triangle arrangement.

Draw an imaginary line connecting the points. Do not place flowers outside this line. The spacing between these flowers should be greater than that nearer the focal area, which is where the lines of the design meet at a 90° angle. Place flowers closely together at this point. Note the staggered placement of flowers which gives depth to the arrangement (Figure 12-25).

Step 5. The side of a right triangle arrangement is visible so it must contain flowers also. Keep flowers close to the main stem to maintain the right angle design (Figure 12-26).

Step 6. Green the arrangement placing foliage behind and in front of flowers (Figure 12-27). Remember to

FIGURE 12-25

Add mass flowers along the lines of the triangular pattern.

FIGURE 12-26

Place flowers along the side of the arrangement.

FIGURE 12-27

The completed design.

green the back of the arrangement. Add filler flowers as needed.

Step 7. Evaluate the arrangement using the flower arrangement rating scale in Appendix G.

Student Activities

1. Complete each of the arrangements described in this unit according to the patterns given. After you have practiced using the patterns, use your own creativity in designing the same arrangements without following the exact pattern.

2. Examine florist trade magazines to find examples of triangular arrangements. Try copying some of the designs.

3. Plan a small flower show in your class using triangular designs, and offer prizes to the winner.

4. Invite a florist to your class to discuss and illustrate triangular designs used in his/her flower shop.

5. Sketch the design form for each of the triangular designs and show the radiating line movement of each design.

Self-Evaluation

A. Evaluate the arrangements using the Flower Arrangement Rating Scale in Appendix G.

B. Short Answer Questions

1. Explain how rhythm is used in designing triangular arrangements.

2. What are some possible uses for triangular designs?

3. How does the isosceles triangle design differ from the equilateral triangle?

4. Why is the centerpiece design appropriate as a table centerpiece?

5. How does a scalene triangle arrangement differ from an asymmetrical triangle arrangement?

Line Arrangements

Terms to Know

negative space

OBJECTIVE

To design linear arrangements.

Competencies to Be Developed

After completing this unit, you should be able to:

- construct an inverted-T arrangement.
- construct an L-pattern arrangement.
- construct a vertical arrangement.
- construct a crescent arrangement.
- construct a Hogarth curve arrangement.
- construct a contemporary freestyle arrangement.

Introduction

Line arrangements lead the eye along an obvious path in the design and keep the eye in continuous motion. The line must never be broken if line movement is to be continued. In this unit, three forms of line arrangements will be studied: straight line, curvilinear, and contemporary freestyle designs.

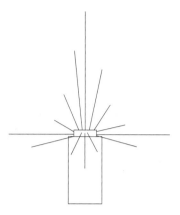

FIGURE 13-1

Design pattern and line movement for inverted-T arrangement.

INVERTED-T ARRANGEMENT

The **inverted-T arrangement** is a variation of the equilateral triangle. The points of this design actually form an equilateral triangle. However, the inverted-T arrangement makes use of **negative space**, or areas without flowers. You'll notice such spaces between the side points and the top of the arrangement (Figure 13-1).

Constructing an Inverted-T Arrangement

Step 1. Select materials:

low container
5 snapdragons
4 or 5 carnations
2 stems of pompons
leatherleaf stems
floral foam

Step 2. Prepare the container. Secure saturated foam into the vase with anchor tape. Leave 1 inch above the rim of the container.

Step 3. Select three of the least open line flowers to establish the points of the design. If line flowers are not available, create a trunk line with mass flowers. The center flower should be placed at the back center of the foam leaning back slightly. These three flowers should form an equilateral triangle (Figure 13-2).

FIGURE 13-2

Skeleton flowers for the inverted-T arrangement.

FIGURE 13-3

Placement of flowers four and five.

Step 4. Place two additional snapdragons to the left and right of the center stem as shown in Figure 13-3.

Step 5. Add mass flowers. The mass flowers may be added in a symmetrical fashion or the flowers may form a graceful sweeping curve within the boundaries of the design pattern. This gives the design a less formal look and creates an interesting line movement through the arrangement. Begin by placing carnations below the left center snapdragon and curve downward across the focal area (Figure 13-4).

Step 6. Add pompon chrysanthemums in the open spaces and green the arrangement by placing sprigs of leatherleaf in front of the flower stems but below the flower heads. Add foliage to the back of the arrangement as well. Insert small sprigs of statice. Place some of these deep in the arrangement to create depth in the design (Figure 13-5).

Step 7. Evaluate the design using the rating scale in Appendix G.

FIGURE 13-4
Add carnations in a graceful, sweeping curve.

FIGURE 13-5
The completed
inverted-T
arrangement.

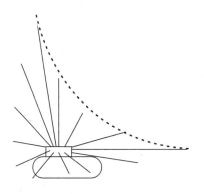

FIGURE 13-6
A modified L-pattern design.

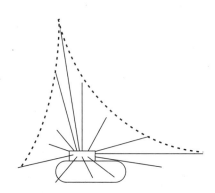

FIGURE 13-7
A three-legged L-pattern design.

L-PATTERN ARRANGEMENT

The **L-pattern arrangement** is very similar to the right triangle arrangement, but it is more linear since the area between the two major points remains unfilled. This gives the L-pattern a lighter, more stylized appearance. In a true L-pattern, the vertical line is straight up and the horizontal line is parallel with the table, creating a 90° angle where the two lines meet. In a modification of the L-pattern, the vertical line is slanted to the left and the horizontal line sweeps downward (Figure 13-6). A third line may be added to create a three-legged L-pattern (Figure 13-7).

Constructing an L-Pattern Arrangement

Step 1. Select materials:

2 gladioli
9 carnations
1 stem of statice
leatherleaf
low container
floral foam

Step 2. Prepare the container. Position one-third of a block of saturated floral foam to the left side of a low flat container. Secure with anchor tape or anchor pins.

Step 3. Establish the height of the arrangement with a gladiolus that has the bottom floret removed. Insert the stalk in the back left corner of the foam. The horizontal line may be three-fourths the length of the vertical line. Remove the bottom two florets from the second gladiolus and save these for later use. Position the second gladiolus parallel to the table so that the vertical and horizontal lines meet at a 90° angle (Figure 13-8).

Step 4. Add the gladiolus florets with a short piece of stem to the arrangement in the manner shown in Figure 13-9. Begin with the stem with two florets. Angle this stem forward and to the left to give depth to the arrangement. The second flower is placed at a 45° angle between the two main lines. This flower should never be more than one-quarter the length of the vertical line or the L-pattern will be destroyed. Using the gladiolus florets in this manner will help to unify the arrangement.

Step 5. Add the carnations in a rhythmic fashion along the two lines of the arrangement (Figure 13-10).

Step 6. Green the arrangement by placing sprigs of foliage in front of the flower stems but below the flower

FIGURE 13-8
The skeleton pattern of the L-arrangement.

FIGURE 13-9
Use gladiolus florets to unify the arrangement.

FIGURE 13-10

Add carnations along the skeleton lines of the arrangement.

heads. Also green the back of the arrangement. Add small sprigs of statice in the spaces between the carnations (Figure 13-11).

Step 7. Evaluate the arrangement using the flower arrangement rating scale presented in Unit 2.

FIGURE 13-11

The completed L-arrangement.

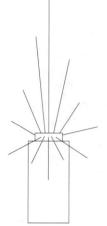

FIGURE 13-12

The design pattern for the vertical design.

THE VERTICAL ARRANGEMENT

The **vertical arrangement** has a very strong line and makes a bold statement. The eye tends to continuously move up and down the arrangement. The tall, slender vertical arrangements are excellent for hospitals and nursing homes where space for flowers is limited. It also attracts the eye while using a limited number of flowers (Figure 13-12).

Constructing a Vertical Arrangement

Step 1.   Select materials:

> a tall cylindrical container
> 1/4 block of floral foam
> line foliage such as camellia, myrtle, or
>   Scotch broom
> 6 carnations
> 6 blooms of pompons

Step 2.   Prepare the container. Secure saturated foam in the vase with anchor tape, leaving 1 inch of foam above the rim of the container.

Step 3.   Insert a tall piece of greenery two to three times the height of the vase into the floral foam at the center back of the foam (Figure 13-13). This piece of greenery establishes the height and central axis of the arrangement.

Step 4. Add carnations. Place the first carnation leaning just to the left of the central axis and below the tip of the foliage. Place a second carnation at the center of the arrangement near the rim of the container. This carnation should face forward and stick out a couple of inches from the container. Place the remaining four carnations in a staggered position as shown in Figure 13-14. Facing the flowers and proper space rhythm are important.

Step 5. Insert pompon chrysanthemums in the open spaces between carnations (Figure 13-15). Flowers should extend only slightly to the right or left of the vase.

FIGURE 13-13
Establish the height of the vertical design.

FIGURE 13-14
Placement of carnations for the vertical design.

FIGURE 13-15
Place pompons in open spaces between carnations.

Step 6. Place sprigs of foliage in the open spaces below and between the flowers (Figure 13-16). Do not forget to green the back of the arrangement also.

Step 7. Evaluate the arrangement using the rating scale in Appendix G.

FIGURE 13-16

The completed vertical arrangement.

CRESCENT ARRANGEMENTS

The **crescent design** is a portion of a circle like the moon in its first quarter. This design is sometimes compared to animals' horns. The tips of each horn are pointed and curve down to join at the thicker center. In the crescent design, one horn is longer than the other, with the upper one twice the length of the lower horn. The focal area for this arrangement is located where the two horns meet.

The crescent design may be viewed from one or both sides. If the arrangement is to be viewed from both sides, then each side of the horns would be finished with flowers. If the arrangement is to be viewed from only one side, some of the flowers should face backward to give the arrangement depth.

Constructing a Crescent Arrangement

Step 1. Select materials:

       container
       floral foam
       6 stems of eucalyptus
       1/2 bunch of pixie carnations
       2 stems of statice
       leatherleaf stems

Step 2. Prepare the container. The crescent is easier to build in a shallow bowl, but a pedestal container is also suitable. Place the foam off-center if an oblong container is used. Use a four-pronged anchor pin to hold the foam if anchor tape would be difficult to hide.

Step 3. Select a curved stem of eucalyptus. Place this stem in the foam at the rear left corner. This stem of eucalyptus should be twice the height of the lower stem. Select a shorter stem of eucalyptus and insert it at the front right corner of the foam at a slightly forward angle. This placement will cause the line of the design to come diagonally across the foam. The stems of the eucalyptus should be placed so that the two would form a circle if their lines were continued (Figure 13-17).

FIGURE 13-17

Use eucalyptus to form the crescent pattern of the crescent arrangement.

FIGURE 13-18

Add miniature carnations following the lines of the crescent.

Step 4. Position additional curved stems of eucalyptus in a staggered fashion to emphasize the curve of the design.

Step 5. Add the pixie carnations, following the line established by the eucalyptus. Use buds at the tips of the crescent line and gradually increase their sizes as the focal area is approached (Figure 13-18). Using a 20-gauge wire, wire the stems of the carnations so that the stems can be gently curved. Straight-stemmed flowers may be placed in a staggered line to conform to the outline of the crescent design. Face some flowers toward the back in the focal area or finish both sides of the arrangement. As the focal area is approached, the curve becomes widest.

Step 6. Add leatherleaf to the arrangement to hide mechanics and give depth to the arrangement. Insert small sprigs of statice to fill voids and further develop depth (Figure 13-19).

FIGURE 13-19

The completed crescent arrangement.

Step 7. Check your work and evaluate the design using the flower arrangement rating scale in Appendix G.

THE HOGARTH CURVE ARRANGEMENT

The **Hogarth curve**, also called the **S-curve**, is named after the English painter, William Hogarth (1697–1764). Whereas the crescent is derived from a single circle, the S-curve comes from two circles (Figure 13-20). There are three types of S-curve arrangements: the **classic vertical**, the **diagonal S-curve**, and the **horizontal S-curve** (Figure 13-21).

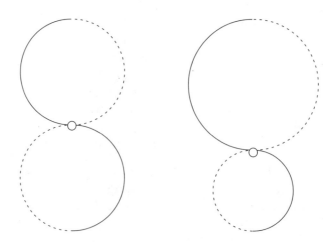

FIGURE 13-20

The Hogarth curve.

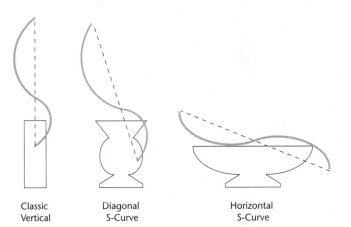

Classic
Vertical

Diagonal
S-Curve

Horizontal
S-Curve

FIGURE 13-21

Types of S-curve arrangements.

Constructing a Hogarth Curve Arrangement

Step 1. Select materials:

container
floral foam
3 carnations
3 stems of miniature carnations
leatherleaf stems
Scotch broom stems

Step 2. Prepare the container. Select a tall pedestaled, cylindrical, or oval vase for this arrangement. Because the Hogarth curve is one of the most elegant of all flower forms, select an elegant container. Cut saturated floral foam to fit the container and secure with anchor tape.

Step 3. Select a piece of Scotch broom twice the height of the container. Work it between your fingers to form a gentle curve (Figure 13-22). Insert this stem in the back left corner of the foam. The tip of this stem should be directly over center in a classic vertical S-curve and to the left of center in a diagonal S-curve. Select a second piece of Scotch broom about half the length of the first. Work the stem between your fingers to form a curve. Place this piece in the right front corner of the foam so that it bends forward and downward (Figure 13-23).

FIGURE 13-22

Work Scotch broom between your fingers to form a curve.

FIGURE 13-23

Place the Scotch broom to form the S-pattern of the design.

FIGURE 13-24

Place flowers along the lines of the S-curve.

Step 4. Add additional pieces of Scotch broom to form the S-curve. Place mass flowers as shown in Figure 13-24. Begin with the miniature carnations using buds at the upper and lower lines of the arrangement. Place the larger carnations near the focal area.

Step 5. Add additional miniature carnations following the S-curve line of the design (Figure 13-25). Turn some flowers to the back to give the arrangement depth and a three-dimensional look.

Step 6. Green the arrangement using small sprigs of leatherleaf to cover mechanics and fill empty spaces (Figure 13-26).

Step 7. Check your work and evaluate the design using the flower arrangement rating scale in Appendix G.

FIGURE 13-25
Add additional miniature carnations.

FIGURE 13-26
The completed Hogarth curve arrangement.

CONTEMPORARY FREESTYLE ARRANGEMENTS

Contemporary freestyle arrangement (see color insert) gives designers the opportunity to express their creativity. The name suggests freedom, but basic principles of design still apply. Geometric forms can be used in contemporary freestyle and great use is made of lines—not just vertical and horizontal lines, but spirals, curves, and diagonals. Lines are very prominent in contemporary freestyle arrangements.

Designers have described contemporary freestyle as personalizing what you have learned about different designs. This style involves placing flowers into a pleasing composition without falling into a particular design style. Designers may mix several styles as long as one is not dominant.

As a beginning designer, contemporary freestyle may seem difficult. Begin by deviating from the rules you have learned in creating other designs. Exaggerate the use of line or combine lines. Make greater use of negative space.

Constructing a Contemporary Freestyle Arrangement

Step 1.   Select materials:

> 3 hybrid tea roses, irises, or carnations
> 1 or 2 lotus pods
> 4 sprigs of pittosporum or other linear foliage
> 6 fronds of Boston fern
> 9 galax leaves
> beargrass
> 6 clumps of reindeer moss
> floral foam
> container (low bowl or other contemporary container)

Step 2.   Prepare the container. Secure saturated foam in the container so that approximately 1 inch of foam sticks out of the container. If anchor tape would be visible, use an anchor pin to secure the foam.

Step 3.   Insert the three roses, or other flowers, into the foam in the following manner. If you are using a low container, cut the longest rose about 21 inches long. Place this rose into the foam slightly to the rear and left of center and leaning gently to the left. The second rose should be about 17 inches long and inserted slightly behind and to the right of center, leaning slightly to the right. Cut the third rose about 12 inches long and place it slightly forward of center and leaning slightly forward and to the left. Figure 13-27 shows the placement of the three roses, which should appear to come from a central point in the floral foam. Leave the foliage on the roses if it is attractive, or remove the leaves from the stem if it does not look good.

Step 4. Place four or five sprigs of pittosporum into the arrangement following the lines of the roses but well below the heads of the flowers (Figure 13-28).

Step 5. Add the fern fronds and beargrasses so that they appear to radiate from the center of the arrangement (Figure 13-29).

FIGURE 13-27

Placement of roses in the contemporary freestyle arrangement.

FIGURE 13-28

Add sprigs of pittosporum to the arrangement.

FIGURE 13-29

Addition of fern fronds and beargrass.

Step 6. Insert a lotus pod at a 45° angle into the front right of the arrangement. The lotus pod should be at or near the rim of the container. If the back of the arrangement will be visible, add a second pod there. Add galax leaves around the base of the arrangement (Figure 13-30).

Step 7. Small clumps of reindeer moss may be added to the base to cover mechanics and add to the contemporary look of the arrangement. Make a hairpin out of one-half of a 20-gauge wire. Use this hairpin to attach clumps of reindeer moss to the base of the arrangement (Figure 13-31).

Step 8. Step back and evaluate your work (Figure 13-32). Make any adjustments you feel are needed. Use the rating scale in Appendix G to evaluate the arrangement.

FIGURE 13-30
Add lotus pods and galax leaves.

FIGURE 13-31
Use a wire hairpin to
add clumps of reindeer
moss.

FIGURE 13-32
The completed
contemporary freestyle
arrangement.

Student Activities

1. Complete each of the arrangements described in this unit according to the pattern given. After you have practiced using the patterns, use your own creativity in designing the same arrangements. Use different flowers and foliage with different containers.

2. Examine florist trade magazines to find examples of line arrangements.

3. Plan a small flower show in your class using line arrangement. Offer prizes to the winner.

4. Invite a florist to your class to illustrate one or more line designs and discuss how they are used in the flower shop.

Self-Evaluation

A. Use the rating scale in Appendix G to evaluate the completed designs.

B. Sentence Completion

1. The inverted-T arrangement is a variation of the _____ triangle.

2. _____ space refers to areas without flowers.

3. The L-pattern arrangement is very similar to the _____ arrangement.

4. The L-pattern and the _____ arrangement make use of large areas of negative space.

5. In a _____ arrangement, the eye tends to continuously move up and down the arrangement.

6. The _____ design is sometimes compared to animal's horns.

7. The S-curve is also called the _____ curve.

8. Contemporary freestyle arrangements give designers the opportunity to express greater _____.

C. Short Answer Questions

1. Why is the vertical arrangement an excellent choice for hospitals and nursing homes?

2. Name and sketch three types of S-curve arrangements.

3. What are the characteristics of the contemporary free-style design?

Constructing Wreaths

greening pins

OBJECTIVE

To design wreaths using evergreens and permanent materials.

Competencies to Be Developed

After completing this unit, you should be able to:

- construct grapevine wreaths using two different methods.
- decorate a grapevine wreath using evergreen and dried materials.
- construct an evergreen wreath using a Styrofoam wreath ring.
- construct an evergreen wreath using a circular wire frame.

Introduction

Wreaths date back to the Greek period where they were worn during special events. Wreaths were also presented to outstanding athletes to symbolize victory and dedication.

Today wreaths are a common form of permanent and seasonal design in our homes and businesses. A wide variety of sizes and shapes are available made from many different materials, including pinecones, straw, grapevines, birch branches, willow, wire, and Styrofoam. These wreath forms may be decorated with a wide variety of evergreen, silk and dried materials. In this unit, we explore a variety of methods used to construct and decorate wreaths.

GRAPEVINE WREATHS

First, two methods of constructing grapevine wreaths are explained.

Grapevine Wreath: Method One

Step 1.   Select materials:

> grapevines (wild vines are best)
> one chenille stem
> paper twist
> a wreath form
> wooden picks
> greening pins

Step 2.   Begin the wreath by bending a vine into a circle the size of the desired wreath. Wire the vine into place (Figure 14-1).

Step 3.   Continue the wreath by twisting the loose end of the vine in and out around the circle. Secure the end of the vine by tucking it among the strands of the circle (Figure 14-2).

FIGURE 14-1

Beginning a grapevine wreath.

FIGURE 14-2

Continue the wreath by twisting vines in and out around the wreath.

FIGURE 14-3

Add a bow to complete the grapevine wreath.

Step 4. Add additional vines by tucking and twisting vine ends into the circle as instructed in step 3.

Step 5. Continue adding vines until the desired size of wreath is reached.

Step 6. Make a bow of paper twist or other material with 12-inch streamers. Tie the bow with a 22-gauge wire. Use the ends of the wire holding the bow to attach it to the wreath (Figure 14-3).

Step 7. Make a hanger for the wreath out of a chenille stem. Twist the chenille stem around several strands of vine so that it will be secure.

Step 8. Check and evaluate your work.

Grapevine Wreath: Method Two

Step 1. Select materials:

> grapevines
> raffia
> lotus pods
> poppy seedheads
> glycerinized eucalyptus
> German statice

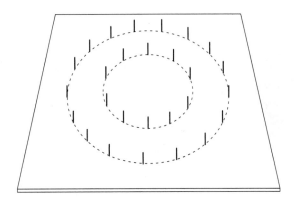

FIGURE 14-4

A homemade wreath form.

FIGURE 14-5

Tie a wire around one side of the wreath.

one 4-inch square of Styrofoam
Spanish moss or sheet moss
homemade wreath form

Step 2. Make a wreath form out of plywood and finishing nails. Draw circles the size of the desired wreath and place nails every few inches along the circle (Figure 14-4).

Step 3. Place grapevines inside the circle of nails. If you have difficulty keeping vines in place, weave them through the finishing nails. Add vines until the wreath is the desired size.

Step 4. Secure the wreath by wiring one side of the wreath with a 22-gauge wire (Figure 14-5).

Step 5. Weave a long slender vine under and over the vines in the wreath form (Figure 14-6). This will hold the vines in place. Secure the ends of this vine underneath the wire. The wreath can now be lifted off the form.

Step 6. Wire a 4-inch square of Styrofoam to the wreath (Figure 14-7). Place wooden picks on the top of the Styrofoam to keep the wires from cutting the Styrofoam. Grapevine and willow wreaths have an open weave. This allows floral materials to be attached with glue. Use this method if you prefer glue.

FIGURE 14-6
Weave a long slender vine around the wreath.

FIGURE 14-7
Add a small piece of Styrofoam to the wreath.

Step 7. Cover the Styrofoam with Spanish moss or sheet moss and secure with **greening pins**. These are wire pins with a U shape that are used as fastening devices (Figure 14-8). The Styrofoam will allow you to add dried or permanent flowers to the wreath.

FIGURE 14-8
A greening pin.

FIGURE 14-9
The decorated grapevine wreath.

FIGURE 14-10

High quality ribbons and silk flowers give the grapevine wreath a more formal look.

Step 8. Select a variety of dried materials to decorate the wreath (Figure 14-9). The same materials could be placed in a diagonal line or a crescent design. A more formal look could be achieved by using more expensive ribbons and silk flowers (Figure 14-10).

Step 9. Check your work and evaluate the wreath.

EVERGREEN WREATH

The evergreen wreath is one of the most popular decorations for the Christmas season. Wreaths are often placed on doors, over the fireplace mantle, and on walls. A simple evergreen wreath with a red velvet bow is a traditional favorite. However, wreaths can be elaborately decorated with fruit, pine cones, nuts, berries, and ribbons of all kinds.

Evergreen wreaths can be constructed using a variety of methods. Evergreen boughs can be wired to a single hoop of wire or attached to a straw or a Styrofoam wreath form with greening pins or steel picks. The Smithers Oasis Company also makes a plastic wreath form filled with floral foam to keep evergreen materials fresher for a longer period of time. For this activity, we will be using a Styrofoam wreath and greening pins.

Evergreen Wreath: Method One

Step 1. Select materials:

> 14-inch or 16-inch wreath form wrapped in green plastic
> greening pins
> 6-inch to 8-inch evergreen boughs
> number 40 red velvet ribbon

Step 2. Bunch two to four pieces of evergreen together on the form and secure them with a greening pin (Figure 14-11).

Step 3. Repeat step 2 but move to the inside, then to the outside of the wreath.

Step 4. Continue to add clusters of evergreens by overlapping the first layer (Figure 14-12).

FIGURE 14-11

Add evergreen boughs to the wreath with greening pins.

FIGURE 14-12

Continue adding evergreens by overlapping the first layer.

FIGURE 14-13

Add evergreens until the wreath form is completely covered.

Step 5. Continue around the wreath until the wreath form is completely covered. Add additional greens where needed. Check the inside and outside of the wreath form for even evergreen coverage. If any pieces stick out, pin or clip them so they're even with the rest of the greens (Figure 14-13).

Step 6. Make a bow out of number 40 red velvet ribbon, leaving 12-inch streamers. Attach the bow to a wooden pick. The bow can be placed on the wreath at the top center, bottom center, or to the left of top center. For this example, the bow will be placed to the left of top center. Attach small pieces of greenery to a wooden or steel pick and work these into the loops of the bow (Figure 14-14).

Step 7. The wreath can be used as it is or it can be further decorated with the addition of pine cones, berries, or any number of accessories. Pine cones can be wired (Figure 14-15) and attached to the wreath ring. Holly berries or nandina berries can be

FIGURE 14-14

Add a bow and work small sprigs of greenery into the bow.

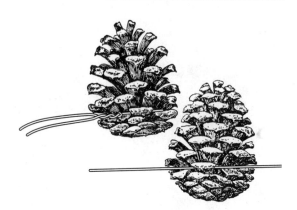

FIGURE 14-15

Wire pinecones to attach them to a wreath.

attached to a steel pick for insertion. Figure 14-16 shows the same wreath with the addition of berries. For a contemporary look, use white velvet ribbon and dip nandina berries in white paint before adding them to the wreath.

Step 8. Make a hanger for the wreath out of one-half of a chenille stem. Fold a chenille stem in half and attach a steel pick. Insert the hanger into the top of the wreath on the back side (Figure 14-17).

Step 9. Check your work and evaluate the wreath.

A second method of constructing evergreen wreaths is done by using wire forms. Figure 14-18 shows three different kinds of wire forms that can be used.

When using the metal contour, weave bundles of greens between the wire mesh and secure by wrapping with paddle wire that has been attached to the frame. The second bundle is positioned with the tips overlapping the first bundle approximately half way. This process of binding bundles to the frame is continued all the way around the wreath.

FIGURE 14-16
Add berries to decorate the wreath.

FIGURE 14-17
Attach a hanger to the back of the wreath.

A wire ring with clamps is another way of constructing an evergreen wreath. In this method, evergreen boughs are placed on the ring inside the clamps. A machine (Figure 14-19) operated by a foot pedal closes the clamps one at a time as the wreath is constructed. As new boughs are added, they cover the wire clamp holding the preceding boughs. Proceed around the wreath ring until the wreath is completed.

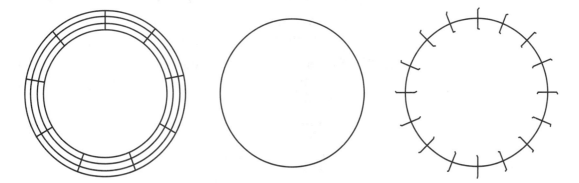

FIGURE 14-18
Three different kinds of wire forms used to construct evergreen wreaths; the metal contour, the wire ring, and the wire ring with clamps.

FIGURE 14-19

A machine operated by a foot pedal is used to close the clamps on a metal wreath ring.

The circular wire frame is the third type of wire wreath frame. The frames may be purchased in various sizes or constructed from stiff wire that is formed into circles. Evergreen wreaths are constructed on the wire frame by tying the greenery with a continuous strand of paddle wire. This method will be used as a class project.

Evergreen Wreath: Method Two

Step 1.   Select materials:

        wire wreath frame
        evergreen boughs cut in 6-inch to 8-inch
           lengths
        number 40 velvet ribbon
        paddle wire
        small pine cones

Step 2.   Cut boughs of evergreen foliage in 6- to 8-inch lengths. Spruce, pine, fir, leyland cypress, and cedar have attractive foliage for Christmas wreaths.

Step 3.   Gather the cut evergreens into a small bundle and place the bundle onto the center of the wire frame and tie it by pulling the paddle wire twice around the stems (Figure 14-20).

Step 4.   Place additional bundles of evergreens slightly below and to each side of the first bundle, and tie these with the paddle wire, (Figure 14-21). Tie the

FIGURE 14-20
Attach bundles of evergreen to a wire using paddle wire.

bundles firmly approximately one-third to halfway from the bottom end of the bundles.

Step 5. Continue placing bundles of evergreens onto the wire frame in the pattern described in step 4 until the wire frame is filled. Each bundle should overlap the previous bundle about halfway and cover the paddle wire. Place the bundles thicker and closer

FIGURE 14-21
Place additional bundles of evergreen to each side of the first bundle and secure with paddle wire.

together for a full heavy wreath. Once the circle of foliage is completed, cut the paddle wire and tie it to the wreath frame.

Step 6. Attach a bow at the point where the circle of foliage meets.

Step 7. Wire small pine cones such as Virginia pine as shown in Figure 14-15. After the wires are attached to the pine cone, push the ends of the wire through the foliage and secure at the back. Push all wire ends back into the wreath. Vary the placement of the pine cones so that some are in the center of the wreath and some are to the right and left of the center. Trim excess foliage and the wreath is complete (Figure 14-22).

Step 8. Check your work and evaluate the wreath.

FIGURE 14-22

The completed evergreen wreath.

Accessories other than pine cones may be used to decorate the wreath. Also popular are sprigs of holly with berries or other clusters of berries such as nandina. Christmas balls and fruit are also popular. For a more contemporary look, use a gold-colored bow and attach lemons around the wreath.

PERMANENT EVERGREEN WREATHS

Wreaths made from artificial evergreen are sometimes desirable, especially in warm climates where live foliage may dry and begin to shed. Wreaths made from artificial evergreen can also be stored and used for a number of years. Businesses also prefer permanent evergreen wreaths because the wreaths often hang for a long period of time during the Christmas season.

Constructing a Permanent Evergreen Wreath

Step 1. Select materials:

> Styrofoam or straw wreath form
> artificial evergreen swags
> ribbon
> greening pins

Step 2. Begin the wreath by securing the end of a swag to the center of the wreath form. Continue attaching the swag to the form with greening pins every 6 inches until you have circled the wreath form. Cut off any excess swag.

Step 3. Repeat step 2 on the inside and the outside of the wreath form.

Step 4. Add a bow and a hanger and the wreath is complete (Figure 14-23).

Step 5. Check your work and evaluate the completed design. Add whatever accessories you desire to decorate the wreath or add springs of live foliage to give the wreath a more natural appearance. To add the live springs, simply press them into the artificial foliage or place the ends of the stems underneath the wire forming the swag (Figure 14-24). If the live

FIGURE 14-23

A permanent evergreen wreath.

sprigs begin to shed, they can be easily replaced with minimal effort. This works particularly well for people who live in warm climates.

EUCALYPTUS WREATH

Preserved eucalyptus is a popular foliage for decorating grapevine wreaths and may be used as the main foliage in a wreath. We will not do a step-by-step process, but once you

FIGURE 14-24

Add sprigs of live evergreen to an artificial wreath for a natural appearance that can be easily changed.

have mastered the other wreaths, eucalyptus wreaths will not be difficult to construct.

Cut preserved eucalyptus into 8-inch to 10-inch lengths and place in bundles of three to four stems. Place a steel pick on the end of each bundle. Secure the bundles into a Styrofoam or straw wreath form until the wreath ring has been covered and is thick and full. Add ribbons and whatever accessories you desire and the wreath is complete (Figure 14-25).

FIGURE 14-25
A eucalyptus wreath

Student Activities

1. Complete one or more of the wreaths described in this unit. Gather your own foliage to keep the cost of the wreaths to a minimum.
2. Make wreaths to sell to the school faculty or to other students.
3. Gather pictures of wreaths from magazines and catalogs and make a class scrapbook of the designs for the class to use to obtain wreath-decorating ideas.

Self-Evaluation

Short Answer Questions

1. Name several materials from which wreaths are made.

2. Describe two ways that accessories may be added to a grapevine wreath.

3. Explain how a homemade grapevine wreath form can be constructed.

4. What are greening pins?

5. List several types of evergreen available in your area that can be used to make evergreen wreaths.

6. List three types of wire wreath forms.

7. What are some of the most common materials used to decorate evergreen wreaths?

8. Describe one method that can be used to make a hanger for a Styrofoam or straw wreath.

9. Explain how to make an artificial evergreen wreath look more natural.

Holiday Arrangements

Terms to Know

curling ribbon
heat sealer
helium
Hi Float
latex
Mylar balloons
regulator
topiary

OBJECTIVE

To design floral arrangements for holiday sales.

Competencies to Be Developed

After completing this unit, you should be able to:

- make an arrangement for Valentine's Day.
- make an arrangement for Easter.
- make an arrangement for Mother's Day.
- make an arrangement for Thanksgiving.
- make an evergreen wreath or topiary for Christmas.
- make an arrangement for Christmas.
- make a holiday arrangement using balloons.

Introduction

Peak sales in the retail florist business occur during the holidays. Figure 15-1 lists holidays of importance to the retail florist. The major sales periods occur during the four holidays of Christmas, Valentine's Day, Easter, and Mother's Day.

During the holidays, retail florists do the largest volume of business in a very short time. As is common in the retail flower shop business, the florist must plan well and work long hours. The industry depends heavily on these peak

HOLIDAYS	DATE
Valentine's Day	February 14
St. Patrick's Day	March 17
Easter	Variable Dates
Secretaries' Day	4th Wednesday of April
Mother's Day	2nd Sunday of May
Memorial Day	Last Monday of May
Father's Day	3rd Sunday of June
Independence Day	July 4
Grandparents' Day	1st Sunday of Sept.
Bosses' Day	October 16
Sweetest Day	3rd Saturday of Oct.
Mother-in-Law's Day	4th Sunday of Oct.
Halloween	October 31
Thanksgiving	4th Thursday in Nov.
Christmas	December 25
New Year's Eve	December 31

FIGURE 15-1
Holiday calendar.

sales to make up for slack periods occurring particularly during the summer months.

The floral designer must be able to construct creative designs for all of the holiday periods. This unit presents design ideas for five of them. The designs are creative but simple and inexpensive to arrange. Many of the materials can be gathered from the home landscape or from the school grounds. Flowers that have been purchased are inexpensive and readily available. The floral design class can sell its arrangements to students and faculty at the school.

VALENTINE'S DAY

The first big holiday of the year occurs at Valentine's Day, traditionally a time of exchanging tokens of love. Men most often purchase flowers for their girlfriends or wives. The red rose, symbolizing love, is the most popular flower for this holiday. Because of the increased demand, the price of roses greatly increases, and many people seek other flowers. Red tulips, carnations, and mixed bouquets with red heart accessories are popular. Corsages and bud vases are ideal for people wishing to send a floral gift without spending a large sum of money. Two ideas for valentine arrangements are presented in this unit. Other suggestions for the holiday include rose bud vases or a corsage attached to a box of candy.

Carnation in a Coke Can

Step 1.   Select materials:

>1 Coke can
>floral tape
>20-gauge wire
>1 stem of myrtle or Scotch broom
>leatherlcaf stems
>1 stem of pittosporum
>3 red or peppermint carnations
>1 stem of baby's breath
>1 yard of number 3 white or red ribbon
>1 red chenille stem

FIGURE 15-2

Place the three carnations in a Coke can as though you were designing a bud vase.

Step 2.   Empty and wash the Coke can and break out the metal tab. This can will serve as the container for our arrangement. Fill the Coke can with warm preservative solution.

Step 3.   Wire three red or peppermint carnations using a 20- or 22-gauge wire. Arrange these in the can as you would carnations for a bud vase. The tallest carnation should be about three times the height of the can. Stagger the other two carnations with two to three inches of space between each one (Figure 15-2).

Step 4.   Add a stem of myrtle or Scotch broom to the back of the carnations to create a strong vertical line. Add one stem of leatherleaf at the back and a piece of pittosporum at the base (Figure 15-3).

Step 5.   Add sprigs of baby's breath to soften the arrangement.

Step 6.   Tie a loose bow of number 3 satin ribbon and add it to the base of the arrangement (Figure 15-4).

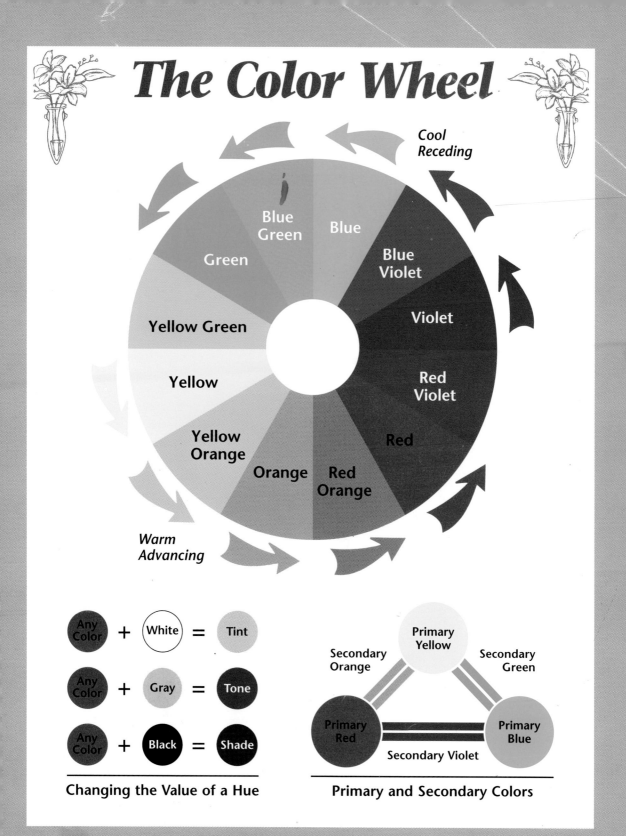

The Color Wheel

Cool
Receding

Blue
Green

Blue

Green

Blue
Violet

Violet

Yellow Green

Red
Violet

Yellow

Yellow
Orange

Red

Orange

Red
Orange

Warm
Advancing

Any Color + White = Tint

Any Color + Gray = Tone

Any Color + Black = Shade

Changing the Value of a Hue

Primary Yellow

Secondary Orange

Secondary Green

Primary Red

Primary Blue

Secondary Violet

Primary and Secondary Colors

Color Harmony

Analogous Color Harmony

Monochromatic
Color Harmony

Triad Color Harmony

Color Harmony

**Complementary
Color Harmony**

**Split Complementary
Color Harmony**

Polychromatic Color Harmony

Shape Arrangements

Scalene Triangle

Contemporary
Freestyle Arrangement

Horizontal Table Arrangement

Holiday Arrangements

Valentine Bud Vase

Christmas Arrangement

Easter Arrangement

Valentine Arrangement

Bouquets

Nosegay

Cascading Bouquet

Arm Bouquet

Corsage attached
to the shoulder
with adhesive tape

Bridal Cascading
Bouquet

Wedding Decorations

Pew
Candelabra

Altar Decorations for a Wedding

Bride's Cake Table

Wedding Reception
Food Table

Sympathy Flowers

Sympathy Wreath

Casket Cover

Sympathy Basket

Sympathy Heart

FIGURE 15-3
Add greenery to the arrangement.

FIGURE 15-4
Tie a bow with heavy wire and add to the arrangement.

FIGURE 15-5
Add hearts made from red chenille stems.

Step 7. Fashion hearts out of two red chenille stems. Attach them with floral tape about one inch apart to a piece of 22-gauge wire. Add the hearts to the arrangement in the space between the top and bottom carnation (Figure 15-5 and the color insert). This completes the arrangement.

Step 8. Check your work and evaluate the arrangement using the rating scale in Appendix G.

Valentine Mug Arrangement

Step 1. Select materials:

> 1 valentine mug
> 18 daisy pompons
> 2 stems of leatherleaf
> 1 stem of baby's breath
> 3 small valentine accessories of any type
> floral foam

FIGURE 15-6

Prepare a small cone-shaped arrangement in a Valentine's Day mug.

FIGURE 15-7

Add valentine accessories to complete the arrangement.

Step 2. Prepare the container. Cut saturated foam to fit the mug and secure it.

Step 3. Prepare a small cone-shaped arrangement in the mug as shown in Figure 15-6.

Step 4. To finish, secure three valentine hearts or other small valentine accessories to a wire or pick and place them in the sides and top of the arrangement (Figure 15-7 and the color insert).

Step 5. Check your work and evaluate the arrangement using the rating scale in Appendix G.

EASTER

Easter is a religious holiday that occurs between early March and late April. It is also the beginning of the spring season. For this reason, spring flowers, such as tulips and daffodils, are popular cut flowers and potted plants. Customers also enjoy potted hydrangeas, but the Easter lily is the most popular potted plant during this season.

Accessories for Easter include Easter baskets, the Easter bunny, colored eggs, chicks, and ducks. Frequently used colors include bright shades of purple, green, yellow, and pink.

Corsages are also popular at Easter with the cymbidium orchid being in great demand. Rose corsages in red and yellow are also popular.

Easter Bucket Arrangement

Step 1. Select materials:

> plastic bucket container
> clearphane film
> 2 daffodil blossoms
> 8 daisy pompons
> 2 stems of leatherleaf
> 1 Easter duck or other stuffed animal
> floral foam

Step 2. Prepare the container. Select a brightly colored plastic bucket for a container. Place floral foam into the container, level with the top of the bucket.

Step 3. Attach the Easter duck to a wooden pick using a 26-gauge wire. Circle the back of the duck with the wire if there is nothing on the bottom which can be used to secure the duck to the wooden pick.

Step 4. Cut an 8-inch square of brightly colored clearphane film. Place this over half of the bucket container and attach the duck to the design by pushing the wooden pick attached to the duck through the film (Figure 15-8). The film will prevent the Easter duck from absorbing water from the floral foam.

Step 5. Arrange the daffodils and pompons in the other half of the bucket as shown in Figure 15-9 and the

FIGURE 15-8
Place the Easter duck on one side of the
arrangement.

FIGURE 15-9
Make an arrangement of daffodils and
pompons in the other half of the Easter bucket.

color insert. Insert a wire up the center of the daf-
fodil stem into the base of the flower. This will
allow you to straighten the head of the flower.

Step 6. Check your work and evaluate the completed design using the rating scale in Appendix G.

Cymbidium Orchid Corsage

Step 1. Select materials:

1 cymbidium orchid
5 small sprigs of ming or sprengeri fern
1-1/2 yards of number 3 picot ribbon
26-gauge wire
corsage box

FIGURE 15-10

Place sprigs of ming at the top of the orchid.

Step 2. Cut the stem of the cymbidium orchid to one inch in length. Cut a 26-gauge wire in half. Insert half of the wire through the stem just below the flower. Insert the other wire through the stem at a right angle to the first. Bend the wires down and wrap them with floral tape.

Step 3. Wire five small sprigs of ming using the wrap-around method.

Step 4. Position one sprig of ming to the upper left area of the corsage. Add a second a little lower on the right side (Figure 15-10).

Step 5. Tie a bow from number 3 picot ribbon (shown in Figure 8-2) and attach it to the stem of the corsage

FIGURE 15-11

Add a bow and additional sprigs of ming.

just below the flower. Check to see that the bow and corsage are in good proportion.

Step 6. Add the three remaining sprigs of ming near the base of the corsage. Work the sprigs into the loops of the bow (Figure 15-11). Cover all wires with florist tape and curl the stem of the corsage.

Step 7. Check your work and then place the corsage in a corsage box and store in the cooler.

MOTHER'S DAY

Mother's Day is celebrated on the second Sunday in May. Honoring mothers, it is one of the busiest holidays of the year. Sentimental flowers similar to those given at Valentine's Day are popular, and roses are always in demand.

Corsages, often made of cymbidium orchids and roses, are a favorite gift item. Traditionally, a red flower is worn as a symbol of love and honor for a living mother, and a white flower is worn as a memorial for a deceased mother.

Because of its lasting quality, the cymbidium orchid corsage is suggested as a class activity for Mother's Day. Another inexpensive suggestion is a rose bowl.

Rose Bowl

Step 1. Select materials:

> 6-inch bubble bowl
> 1 open rose
> 1 1-1/2-inch square of floral foam
> 1 anchor pin
> 1 stem of leatherleaf
> florist clay

Step 2. Use florist clay to secure an anchor pin to the bottom of the bowl.

Step 3. Cut a 1-1/2-inch square of saturated floral foam. Collar the foam with small sprigs of leatherleaf fern by inserting them in a circle around the edge of the foam (Figure 15-12).

FIGURE 15-12
Collar the floral foam with leatherleaf.

FIGURE 15-13

The completed rose bowl.

Step 4. Press the collared foam onto the anchor pin in the bowl.

Step 5. Select a rose that is at least partially open. If necessary, "reflex" the petals by placing the thumb against the middle of the petal and bending the upper part over the thumb at a 45° angle. This causes the rose to fully open.

Step 6. Cut the rose stem to 1-1/2 inches in length.

Step 7. Press the rose stem into the saturated foam until the base of the flower rests on the foam.

Step 8. Add 1 inch of water to the bowl and mist the rose with water.

Step 9. Check your work and evaluate the completed floral piece (Figure 15-13).

THANKSGIVING

Thanksgiving decorations center around the theme of a bountiful harvest. Because the holiday occurs in the fall of the year, dried and permanent flower arrangements and wreaths are popular. See Unit 14 if you would like to construct a wreath for Thanksgiving.

Arrangements for the fall can be made from a number of materials featuring dried grasses, flowers, grains, and nuts. The cornucopia has become a symbol of the bountiful harvest and is a popular decoration. Chrysanthemums are popular flowers in the fall and they are often featured in Thanksgiving arrangements. Fruits of all kinds are also popular. For this Thanksgiving project, we will construct a table arrangement of fresh flowers and fruits for the Thanksgiving table. Do not try to copy the arrangement exactly outlined, but use it as a guide to create your own unique design.

Constructing a Thanksgiving Table Arrangement

Step 1. Select materials:

> a basket with a liner
> 1 bunch of snapdragons or other line flowers such as grasses or seed heads
> 1/2 bunch of leatherleaf
> an assortment of fruit
> 4 fuji or rover chrysanthemums
> floral foam
> 1/2 bunch of pompons

Step 2. Baskets make excellent containers for Thanksgiving arrangements. Select one with a liner (Figure 15-14) and secure floral foam into the liner.

FIGURE 15-14
A basket liner.

FIGURE 15-15

Arrange line flowers as you would for a centerpiece arrangement.

Step 3. Place snapdragons or other line flowers into the floral foam as you would for a centerpiece arrangement (Figure 15-15). See Unit 12 for additional help.

Step 4. Green the arrangement.

Step 5. Drape bananas or grapes across the floral foam and down the side of the basket. Secure with a wooden pick.

FIGURE 15-16

Add fruit along the edge of the basket.

FIGURE 15-17
Add fuji mum to the arrangement.

Step 6. Place additional pieces of fruit on wooden picks and add to the arrangement along the edges of the basket (Figure 15-16).

Step 7. Add fuji mums as shown in Figure 15-17.

Step 8. Add pompons and additional greenery if needed and the arrangement is ready for the Thanksgiving table (Figure 15-18).

FIGURE 15-18
Add pompons and additional greenery if needed.

Step 9. Check your work and evaluate the completed design using the rating scale in Appendix G.

CHRISTMAS

Christmas is the longest and most celebrated of the holidays. No other holiday inspires the use of flowers and decorations as does the Christmas season. It is a time of religious celebrations as well as social and family gatherings. These gatherings inspire the lavish use of decorations both inside and outside the home.

Popular during the Christmas season are decorated potted plants and arrangements of fresh evergreens. Pine, spruce, fir, hemlock, cedar, holly, and juniper are in great demand for wreaths, door swags, and centerpieces. The poinsettia is the most popular potted plant. Traditional colors are red and green, but contemporary colors of silver, gold, and white are also popular. Country colors of blue, rust, and green are frequently used also.

The Christmas season begins early for the retail florist. Wholesalers begin displaying Christmas merchandise early in the fall and retail florists often begin the season with an open house soon after Thanksgiving.

Two ideas for Christmas projects will be featured in this section. Because of the demand for topiaries and centerpieces made from fresh evergreens, these are suggested as class projects.

Evergreen Centerpiece

Step 1. Select materials:

evergreen boughs cut in 8-inch to 12-inch lengths
number 40 velvet ribbon
3 red 18-inch candles
low, flat container
1 block of floral foam
24 4-inch wooden picks

Step 2. Prepare the container. Secure a full block of saturated floral foam into a low, flat oblong container.

Step 3. Insert candles into the floral foam. Attach two 4-inch wooden picks to the base of one candle with

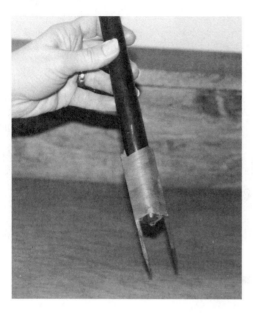

FIGURE 15-19
Attach wooden sticks to base of the candle.

florist tape (Figure 15-19). A plastic candle holder with a pointed base is also available for this purpose. Insert this candle into the center of the foam until the base of the candle rests on the surface of the foam. Insert the other two candles 2 inches into the foam near each end of the container (Figure 15-20). Wooden picks can also be attached to the two outside candles if you have difficulty keeping them in place.

Step 4.   Use the evergreen boughs to create a diamond centerpiece arrangement as previously instructed. If two types of evergreen are used, blend them so that they are uniformly mixed throughout the arrangement (Figure 15-21). Keep the greenery several inches below the candles so they can burn without singeing the evergreens.

Step 5.   Cut twenty 10-inch lengths of number 40 ribbon. Make loops out of twelve of these, and attach them to 4-inch wooden picks. Make streamers out of the remaining eight by making V-shaped cuts into one end of the ribbon and attaching the other end to a wooden pick. Place half of these into each side of the arrangement (Figure 15-22 and the color insert).

FIGURE 15-20
Insert the candles into the floral foam.

FIGURE 15-21

Make a centerpiece arrangement of evergreen boughs.

FIGURE 15-22

Attach ribbon loops to the arrangement.

Step 6. The arrangement can be used just as illustrated or flowers such as red carnations or white daisy pompons can be added (Figure 15-23).

Step 7. Check your work and evaluate the design.

FIGURE 15-23

Add white daisy pompons to the Christmas arrangement.

Topiaries

The **topiary** is a round sphere of flowers and foliages. The shape appears the same from any viewing angle and offers a formal decorating idea in a variety of design styles and sizes.

Topiaries are made from a variety of materials and offer year-round design applications but they are particularly popular during the Christmas season. Two design styles are shown in this unit. One is made of preserved foliages and may be used to decorate the home all through the year. The other topiary is made of live evergreens and has application mainly for the Christmas season.

Constructing a Topiary Using Preserved Foliages

Step 1.　Select materials:

> an 18-inch branch cut at an angle on both ends
> 2 or more small pieces of grapevine 18 inches in length
> sheet moss
> a 4-inch Styrofoam ball
> 1 bag of preserved ming
> 1 bag of preserved asparagus fern
> 1 bag of preserved eucalyptus
> a container
> 2 small squares of 2-inch Styrofoam to fit container

Step 2.　Cut a branch 18 inches long at an angle on both ends. Wrap two or more small stems of grapevine around the branch and tie or glue them so that they end 2 inches from the end of the branch. The first 2 inches of the branch will be inserted into the Styrofoam (Figure 15-24).

Step 3.　Prepare the container by cutting two pieces of Styrofoam the same size as the container. The Styrofoam should fit tightly into the container. Hot glue or use florist putty to attach the Styrofoam to the container.

Step 4.　Wrap the four-inch Styrofoam ball with sheet moss. Attach the moss with greening pins (Figure 15-25).

FIGURE 15-24

Begin the topiary by cutting a branch eighteen inches long and attaching grapevine to it.

FIGURE 15-25

Wrap a four-inch Styrofoam ball with sheet moss.

FIGURE 15-26

Attach the ball to the prepared branch and insert the branch into the Styrofoam container.

Step 5. Put hot glue on the end of the branch and insert it 2 inches into the Styrofoam ball. Hot glue the other end of the branch and insert it 2 inches into the Styrofoam inside the container (Figure 15-26).

Step 6. Cut the ming, eucalyptus, and asparagus fern into 6-inch lengths. Combine several stems and attach them to a steel pick (Figure 15-27).

Step 7. Insert the picked foliage into the Styrofoam ball until it is thick and full (Figure 15-28).

Step 8. Cover the Styrofoam base with sheet moss, using greening pins to hold it in place.

Step 9. Arrange additional foliage at the base and the topiary is complete (Figure 15-29). This topiary can be used year-round but add your favorite Christmas accessories for a festive look over the holidays. Try

FIGURE 15-27

Cut foliages into six-inch lengths and attach small bundles of the foliage to a steel pick.

FIGURE 15-28

Insert the picked foliage into the Styrofoam.

FIGURE 15-29

The completed topiary.

placing fresh cranberries on a small wire and inserting them into the topiary. The cranberries add a holiday look and they can be easily removed after Christmas.

Step 10. Check your work and evaluate the design.

Constructing a Topiary Using Fresh Evergreens

Step 1. Select materials:

evergreens cut into 8-inch to 10-inch lengths
a commode plunger
a small bag of birch bark or other exfoliating bark
hot glue
a 6-inch clay pot
2 small blocks of Styrofoam
gold rope ribbon or other ribbon of your choice.
1 small bunch of dried hydrangea or other dried flower
floral foam

Step 2. The plunger will serve as the reservoir to hold our floral foam and water. Cut the wooden end off to

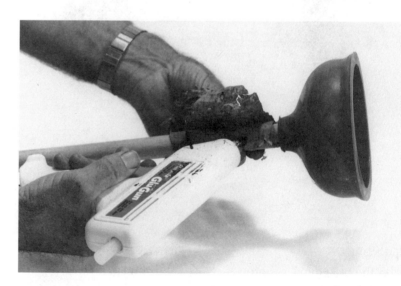

FIGURE 15-30

Hot glue strips of birch bark to cover the plunger.

approximately 18 inches in length. To camouflage its identity glue strips of birch bark over the entire length (Figure 15-30). Remember hot glue burns so be careful.

Step 3. Cut the two blocks of Styrofoam so that they fit tightly into the 6-inch clay pot. Use hot glue or florist putty to hold the Styrofoam in place.

Step 4. Put hot glue onto the end of the plunger and insert it deeply into the Styrofoam.

Step 5. Cut wet floral foam to fit the reservoir of the plunger and tape it in place (Figure 15-31).

Step 6. Insert evergreens into the floral foam, letting sprigs of the greenery drape over and hide the reservoir of the plunger. Continue adding sprigs until the topiary is thick and full (Figure 15-32).

FIGURE 15-31

Tape wet floral foam into the reservoir of the plunger.

FIGURE 15-32

Insert sprigs of evergreen into the floral foam until the topiary is thick and full.

FIGURE 15-33

Add greenery and a sprig of dried hydrangea to the base.

Step 7. Add additional greenery and a sprig of dried hydrangea to decorate the base of the topiary (Figure 15-33).

Step 8. Add a gold rope ribbon bow at the base of the reservoir and the topiary is complete (Figure 15-34). Add other Christmas accessories to give the topiary a more festive look.

MINOR HOLIDAYS

A number of minor holidays such as Secretaries' Day and Grandparents' Day are promoted by florist's associations. See Figure 15-1 for other minor holidays and dates. Created and promoted by the floral industry, each of these minor holidays falls in the yearly calendar when the floral business is

FIGURE 15-34

Add a gold rope ribbon to the base of the reservoir and the topiary is complete.

typically slow, so florists promote them to increase sales. Perhaps the most successful of these minor holidays has been Secretaries' Day.

Floral wire services and wholesalers have designed their own lines of mugs and other specialty containers especially for these holidays. Wire services also promote their own special designs during these holidays.

Talk to your instructor about a special design that your class could construct during any of these minor holidays. Purchase specialty mugs or design your own mugs from plain white mugs purchased from any of the large chain stores. Construct a small cone or vertical design into the mug. Add accessories such as curly ribbon and balloons. Grandparents would love such an arrangement on Grandparents' Day, especially if it were designed by their grandchild. Do similar arrangements for the school secretaries on Secretaries' Day. You can probably get the principal to pay for them.

FIGURE 15-35

A Mylar balloon with candy suckers used as a weight.

HOLIDAY BALLOONS

Balloons are in great demand throughout the year for birthdays, anniversaries, and parties, and they are popular during holidays (Figure 15-35). They are frequently sold individually with a decorated weight, such as a mug or plush animal. They are also sold as part of an arrangement (Figure 15-36). Regardless of how they are used, they are a welcome addition to the items available from the retail flower shop.

Balloons are manufactured from two types of materials: Mylar and latex.

Mylar balloons are produced from a thin metallic film that does not stretch. They are available in many shapes and sizes from 2 inches to 36 inches. For extended periods of time, Mylar will retain either compressed air or **helium**, a nontoxic, nonflammable gas that is lighter than air.

FIGURE 15-36

A floral arrangement containing latex balloons.

Latex balloons are made from latex, a type of rubber, and other chemicals. They stretch when filled with compressed air or helium. Latex balloons are available in many colors and sizes from five inches to four feet. Latex is porous due to very small microscopic holes. Therefore, it does not hold helium or compressed air as long as Mylar. The float time for these balloons can be extended with the use of **Hi Float,** a nontoxic sealant used to treat the interior of latex balloons. Figure 15-37 gives the estimated flying time and helium requirements for various size balloons with and without Hi Float. Helium may be used in latex balloons 9 inches and larger and in all Mylar balloons 18 inches or larger.

BALLOON SIZE	APPROX. CU. FT. OF HELIUM PER BALLOON	AVERAGE FLYING TIME HELIUM ONLY	FLYING TIME HELIUM & HI FLOAT
9" latex	.27	8–16 hours	*
11" latex	.54	12–24 hours	2–5 days
14" latex	1.01	26–40 hours	4–7 days
16" latex	1.51	32–51 hours	7–10 days
40" paddle balloon	14.24	3–5 days	2–4 weeks
3 ft. latex	33.15	**	**
4 ft. latex	65.45	**	**
18" Mylar	.50	3–4 weeks	n/a
36" Mylar	4.00	several months	n/a

*Neither helium nor Hi-Float are appropriate

**Depending upon weather conditions, flying time varies drastically.

FIGURE 15-37

Helium requirements and flying times for balloons with and without Hi Float.

Equipment

Several items of equipment will be needed to inflate balloons. Helium may be purchased in tanks of various sizes from your local welding or oxygen supply company. In addition to the cost of the helium, many suppliers charge a monthly rental fee for the tank.

Please observe all safety precautions when using helium. While helium is a relatively safe gas, direct inhalation can cause dizziness, fainting, suffocation, and possibly death. Helium is also stored in highly pressurized tanks and should be secured in an upright position so that the tank cannot accidentally fall. Tanks can be extremely dangerous if not handled in a safe manner.

A **regulator** is a valve attached to a helium tank to dispense helium into balloons. A regulator may be purchased from a balloon supply company or it may be included in the monthly rental fee for the tank.

An electric **heat sealer** is used to seal the stems of Mylar balloons (Figure 15-38). The sealer should have a thermostat

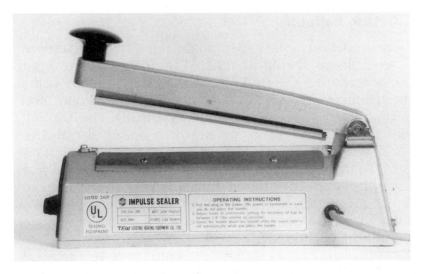

FIGURE 15-38
An electric heat sealer.

to regulate the degrees of heat. Some Mylars are manufactured with a self-sealing valve in the stem and do not require heat sealing or they may be sealed with the use of a clip (Figure 15-39).

Curling ribbon, a special ribbon made from polypropylene that curls when scraped with a knife or scissors, can be used to tie latex or Mylar balloons. Curling ribbon comes in 3/16-inch and 3/8-inch widths and in a variety of colors.

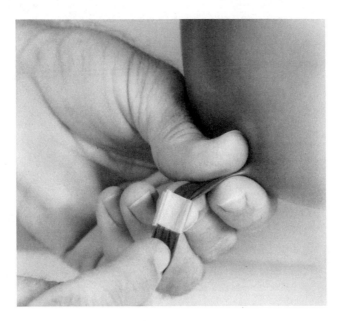

FIGURE 15-39
A balloon clip may be used to seal balloons.

Inflating Balloons

Proper inflation is the key to producing attractive balloons. Latex balloons should be inflated until they are oval, but not pear shaped. Overinflating the latex and then releasing the helium to the correct size will give the balloon more flexibility resulting in fewer balloons bursting. The latex balloon may be hand-tied or sealed with a sealing disc or clip.

A Mylar balloon should be inflated until it is firm but not hard. Remember, Mylar balloons do not stretch, so overinflating can burst the balloon. Inflate slowly. If you use an automatic regulator, it will cut off when full. If the balloon is correctly inflated, you should see about 2 inches of wrinkles around the outside edge.

Mylar balloons are either heat sealed or have a self-sealing valve. The air-filled 2-inch, 4-inch, and 9-inch balloons are sealed and mounted on a cup and stick (Figure 15-40). Mylars 18 inches and larger which are helium-filled are sealed using a heat sealer and a plastic balloon clip added. Tie the balloon with curling ribbon just above the clip.

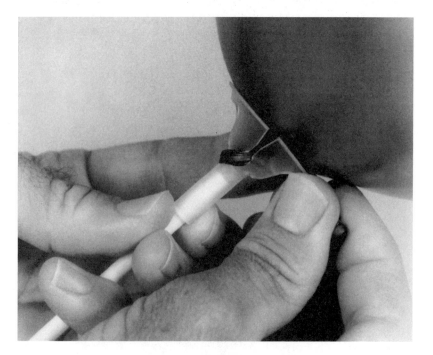

FIGURE 15-40

A balloon may be mounted on a cup and stick.

FIGURE 15-41
A gift basket featuring balloons.

FIGURE 15-42
A bud vase with balloon accessories.

Balloons adapt to any arrangement, whether it is for holiday designs or everyday use (Figures 15-41 and 15-42).

Student Activities

1. Complete each of the holiday arrangements described in this unit. After you have practiced the patterns designed for each holiday, use your own creativity in designing other holiday arrangements.

2. Invite a florist to your class to demonstrate arrangements for the holiday or holidays which occur during your study of this course.

3. Plan to make holiday arrangements to sell to your fellow students or to the faculty of your school.

Self-Evaluation

A. Use the flower arrangement rating scale in Appendix G to evaluate your holiday arrangements.

B. Matching

Match each of the following holidays with the color theme popular for that holiday, see Figure 15-1.

_____ 1. Valentine's Day a. red and white

_____ 2. Easter b. pink, yellow, green, purple

_____ 3. Mother's Day c. black, white, silver

_____ 4. Halloween d. red and green

_____ 5. Christmas e. orange and black

_____ 6. New Year's Eve f. yellow, orange, brown

_____ 7. Thanksgiving g. red, white, blue

_____ 8. Independence Day

C. Sentence Completion

1. _____ is traditionally a time of exchanging tokens of love.

2. The most popular flower for Valentine's Day is the red _____.

3. Two cut flowers that are popular for Easter are the _____ and _____.

4. The _____ is the most popular potted plant at Easter.

5. The _____ is in great demand for corsages at Easter.

6. The _____ is the symbol of a bountiful harvest.

7. The longest and most celebrated of all holidays is _____.

8. The most popular potted plant for the Christmas season is the _____.

9. A properly inflated balloon is _____ shaped.

10. A _____ inch balloon is the smallest Mylar balloon that can be inflated with helium.

D. Short Answer Questions

1. What effect do holidays have on the retail florist industry?

2. What accessories are popular for the Easter season?

3. Make a list of popular potted plants for the Easter holidays.

4. What is the theme for Thanksgiving celebrations and what flowers are popular at this holiday?

5. Make a list of evergreen foliages available for decorating in your area during the Christmas holidays.

6. What is the purpose of using Hi Float in latex balloons?

7. List two ways that latex balloons can be sealed.

8. Calculate the amount of helium needed to fill twenty-five, 18-inch Mylar balloons and fifty 9-inch latex balloons.

Wedding Flowers

OBJECTIVE

To design decorations for a wedding and reception.

Competencies to Be Developed

After completing this unit, you should be able to:

- complete a wedding order form.
- design a bridal bouquet.
- design a bouquet for the bride's attendants.
- identify the areas decorated for a wedding.
- identify decorations used for the reception and rehearsal dinner.

Introduction

Weddings are one of the most interesting and challenging segments of the florist business. In the past, tradition has dictated the planning of flowers and the ceremony. Tradition is still important, but modern brides are planning weddings which reflect their individuality and taste. Generally, couples are older and more educated when they decide to get married and have more definite ideas about what they want in their wedding. The retail florist must recognize the needs of the couple to plan flowers which reflect their individuality.

BRIDAL CONSULTATIONS

The florist may have a number of consultations with the bride. The first interview is usually the most important. This consultation should be conducted at least three months prior to the wedding. These consultations are normally handled by the head designer or the owner/manager of the shop.

Creating the proper setting for the consultation is important. An area should be set aside where the bride and the florist may meet undisturbed (Figure 16-1). The area should be located away from the flow of traffic and the ring of telephones and should contain a table with seating for three or more people. Often the bride is accompanied by her mother and/or the groom. Items such as selection guides, wedding

FIGURE 16-1

A bridal consultation area. *Photo courtesy of M. Dzaman*

forms, examples of ribbons, and photograph albums of wedding work designed by the shop should be present.

The florist must learn as much about the wedding plans as possible. This information will be helpful in planning the wedding flowers and decorations.

Before discussing details for the wedding flowers, the florist must have an idea of what the bride can afford. Encourage the bride to set a budget prior to the consultation or ask probing questions that will help you determine the needs of the bride. The following are suggested as probing questions when asked tactfully. These questions also assist the florist in suggesting flowers for the wedding.

- What type of bridal gown will be worn?
- What type of wedding is being planned?
- How large will the wedding be?
- Will the wedding be formal or informal?
- Where will the service take place and at what time of the day?
- Will the ceremony follow religious customs?
- Where and how large will the reception be?
- Who will be paying for the flowers?

Answers to these questions help establish the size and budget for the wedding. This is a sensitive area and one that the florist must approach tactfully. After obtaining answers to these questions, the florist will be better equipped to make suggestions relative to flower choices and other possible decorations.

In this unit, we will not be able to become knowledgeable about every aspect of the wedding so we will concentrate in those areas that would be most helpful to the beginning designer.

THE WEDDING ORDER FORM

A wedding order form is most helpful in planning the floral decorations for a wedding (Figure 16-2). It lists the majority of items needed for a wedding. These forms, which can be purchased from most wholesale floral suppliers, assist in organizing the wedding order. The form has space for notes on the style and types of flowers used for each part of the wedding. When completed it can serve as a contract between the florist

WEDDING FLOWERS

Bride _____
Address _____
Phone _____
Place of Wedding _____

Date of Wedding _____
Groom _____
Address _____
Phone _____
Time of Wedding _____

Bride's Flowers

Style of Bouquet _____   Bride   Groom

Flowers _____
_____ $ _____
Throw away Bouquet _____ $ _____
Hairpiece _____ $ _____

Attendants' Flowers

		Bride
Honor Attendants _____	Price per Bouquet _____	$ _____
Color of Dress _____	Style of Bouquet _____	
Flowers _____		
Bridesmaids _____	Price per Bouquet _____	$ _____
Color of Dress _____	Style of Bouquet _____	
Flowers _____		
Junior Bridesmaids _____	Price per Bouquet _____	$ _____
Color of Dress _____	Style of Bouquet _____	
Flowers _____		
Flower Girls _____	Price per Bouquet _____	$ _____
Color of Dress _____	Style of Bouquet _____	
Flowers _____		
Hairpiece _____	Price per Hairpiece _____	$ _____
Ringbearers _____	Boutonniere _____	$ _____
Pillow Accessories _____		$ _____

Boutonnieres _____   Price per Boutonniere _____ $ _____

Groom _____   Best Man _____   Groomsmen _____
Ushers _____   Fathers _____   Grandfathers _____
Minister _____   Others _____

Corsages

Bride's Mother _____ $ _____
Groom's Mother _____ $ _____
Grandmothers _____ $ _____
Musicians _____ $ _____
Vocalist _____ $ _____
Guest Book _____ $ _____
Hostesses _____ $ _____
Others _____ $ _____

FIGURE 16-2

A wedding order form.

Church Decorations

Candelabra _____ $ _____

Altar Flowers _____ $ _____

Other Flower Arrangements _____ $ _____

Foliage Decorations _____ $ _____

Aisle or Pew Decorations _____ $ _____

Kneeling Bench _____ $ _____

Other _____ $ _____

Reception

Place _____ Time _____

Bride's Cake Table _____ $ _____

Groom's Cake Table _____ $ _____

Food Tables _____ $ _____

Other Decorations _____ $ _____

Rehearsal Dinner

Date _____

Place _____ Time _____

Decorations _____

_____ $ _____

Bride's Corsage _____ $ _____

Sub total	$ _____	$ _____
Sales Tax	$ _____	$ _____
Total	$ _____	$ _____
Less Deposit	$ _____	$ _____
Balance Due	$ _____	$ _____

Acceptance by the Bride _____

Date _____

Terms _____

FIGURE 16-2

(Continued)

and the bride. Space is available for the bride's signature. This prevents problems in the future. The bride should be given a copy of the form for her reference so she will know exactly what has been ordered and will have an itemized list of her expenses. The form also categorizes the expenses under one column for the bride's flowers and another for the groom's.

BRIDAL AND ATTENDANT BOUQUETS

The bridal and attendant bouquet styles are largely the preference of the bride. However, several factors should be considered when selecting bouquet styles and flowers. The first of these is the style of the wedding gown. A bride who selects a very traditional bridal gown would want to select traditional flowers such as roses and stephanotis. A highly stylized design featuring birds-of-paradise would be inappropriate for this gown.

The size of the bride and bridal attendants also influences the style and size of the bouquets. A petite bride would not want to carry a large, heavy cascading bouquet. A small, airy cascade or a smaller, handheld arrangement would better suit her size. Likewise, a larger bride would want to carry a larger bouquet.

The most frequently ordered bouquet styles for brides and their attendants are the arm bouquet, the colonial nosegay, and the cascading bouquet. Many variations of these can be created by altering the flowers used, the size of the bouquet, and the density of the flowers. In this unit we will make a basic bouquet of each of these styles.

Arm Bouquets

The **arm bouquet**, also called the presentation bouquet, is carried across the forearm (Figure 16-3 and the color insert). It has grown in popularity as a bouquet for both the bride and her attendants, and is designed with the natural stems remaining on the flowers. This offers the advantage of being kept fresh in a vase of water until the wedding. Water picks may also be placed on the individual flowers and removed just before the wedding. This allows the florist to make the bouquets in advance of the wedding. Arm bouquets are constructed with **paddle wire**, a small gauge wire wound around a small wooden paddle.

FIGURE 16-3
An arm bouquet.

Constructing an Arm Bouquet

Step 1. Select materials:

> 3 carnations or roses
> 7 pompons
> myrtle or leatherleaf
> 1 stem of baby's breath
> number 3 ribbon

Step 2. Prepare all flowers and greenery by stripping the foliage from the lower two-thirds of the stem. Wire the carnations for control.

Step 3. Begin with three pieces of strong greenery. Lay the three pieces in a fanlike pattern so that the greenery overlaps.

FIGURE 16-4
Bind the stems of the foliage with paddle wire.

FIGURE 16-5
Add carnations to the bouquet.

Step 4. Bind the stems of the foliage together by wrapping 24-gauge paddle wire around the stems three or four times (Figure 16-4).

Step 5. Add the three carnations at different points on the foliage and bind them with the paddle wire (Figure 16-5).

Step 6. Begin adding the pompons three at a time. Use smaller flowers and buds at the tip of the bouquet and add larger flowers as you near the base. Wrap each group of flowers with the same continuous wire at the same point on the bouquet (Figure 16-6). Do not move up or down the line of flowers.

Step 7. Add additional greenery and filler flowers as needed. Bind the stem three or four more times. Cut the wire and insert the end into the bundle of stems so that it cannot injure the person carrying the bouquet.

Step 8. Use floral tape to tape over the binding point and cover the wire.

Step 9. Trim the ends of the stems to varying lengths. The longest stem below the tape should be about half the length of the bouquet. Stems can vary in length

FIGURE 16-6

Add pompons three at a time.

FIGURE 16-7

The completed arm bouquet.

2 to 3 inches so that the bouquet looks as though you casually gathered a group of flowers and tied a bow around them.

Step 10. At the binding point, add a bow with streamers 8 to 12 inches in length. Tie the wires holding the ribbon at the back of the bouquet and wrap with floral tape. The arm bouquet is now complete (Figure 16-7).

Step 11. Check your work and evaluate the design.

Hand-Tied Bouquets

Hand-tied bouquets are similar to arm bouquets in that they have a very natural appearance and both have the stems remaining on the flowers. The hand-tied bouquet differs in that it generally appears rounded and massed on all sides. The rounded tight cluster of the flowers is the emphasis and individual flowers often lose their focus.

The arm bouquet or presentation bouquet is carried across the arm, whereas the hand-tied bouquet is generally carried in the hand. Because the stems remain on the flowers, the hand-tied bouquet can also be kept fresh by placing it in a vase of water.

Constructing A Hand-Tied Bouquet

Step 1. Select materials:

> 6 springs of greenery (your choice)
> 6 to 8 mass flowers (your choice)
> 6 to 8 sprigs of filler flowers (your choice)
> paddle wire or number 26-gauge wire
> ribbon

Step 2. Prepare all flower and foliage stems by removing the foliage from the bottom half of the stems. This will keep foliage above the binding point of the stems.

Step 3. Begin by holding a single stem of foliage or a flower stem between your thumb and index finger as shown in Figure 16-8. From start to finish, all materials will be held in one hand.

FIGURE 16-8

Begin a hand-tied bouquet by holding a stem of foliage between your thumb and index finger.

FIGURE 16-9

Place a second stem at a slight angle against the first.

Step 4. Place a second stem at a slight angle against the first (Figure 16-9).

Step 5. Continue adding stems of flowers and foliage in a spiral fashion. Allow the tops to angle outward to create a full, round bouquet. The stems will also spiral outward at the base (Figure 16-10).

Step 6. Continue adding stems of flowers and foliages until the bouquet is full and rounded. While still holding the bouquet, bind all the materials together with a wire or string at the point where all stems cross each other (Figure 16-11).

Step 7. The stems can be left irregular for a natural look or cut straight across.

Step 8. Wrap the binding wire with ribbon and tie the ribbon into a loose bow or add a bow attached to a small wooden pick by inserting the wood pick behind the ribbon into the mass of stems. The hand-tied bouquet is now complete (Figure 16-12).

Step 9. Check your work and evaluate the design.

FIGURE 16-10

Continue adding stems of flowers and foliage in a spiral fashion.

FIGURE 16-11

Add flowers and foliage until the bouquet is full and rounded. Bind the stems together at the point where all stems cross.

FIGURE 16-12

Add a bow and the bouquet is complete.

Colonial Nosegay in a Bouquet Holder

Colonial nosegays are handheld bouquets constructed in a circular shape with an attached handle (Figure 16-13). They may be prepared in several different ways. If the bouquet is to be hand-tied, each flower is wired and taped separately, then taped together one flower at a time. This process is time-consuming and increases the cost of the nosegay.

A preferable second method of constructing a colonial nosegay is with a foam bouquet holder. It can be constructed quickly and easily with little concern for flowers falling out of the foam. Because the flowers have a source of water, the nosegay can be made well in advance. Use the following steps to construct a colonial nosegay.

Step 1. Select materials:

miniature carnations
baby's breath
leatherleaf
number 3 satin ribbon
one 4-inch wooden pick

FIGURE 16-13
A colonial nosegay.

Step 2. Soak the foam bouquet holder in preservative water, and place it in a nosegay holder. (A nosegay holder can be constructed by cutting a 1-inch hole in the center of an 8-inch × 8-inch × 1-inch board. Place hot glue on the end of an 18-inch piece of 1-inch PVC pipe and insert it into the hole.)

Step 3. Green the bouquet holder. Place short pieces of leatherleaf all the way around the base of the bouquet holder in a circular pattern (Figure 16-14).

Step 4. Wire the mini-carnations if needed for control.

Step 5. Add four mini-carnations equidistant from each other at the base of the holder. The flowers should extend slightly beyond the foliage (Figure 16-15).

Step 6. Insert one carnation between each of the first flowers to create a circular pattern. The height of the flowers can be varied just slightly to prevent a stiff, fixed look (Figure 16-16).

FIGURE 16-14
Green the base of the bouquet holder.

FIGURE 16-15
Add four mini-carnations.

Step 7. Place from one to three of the largest flowers in the center of the foam holder to create a focal point. The number needed will depend upon how large the flowers are. In this case, use three of the mini-carnations. Vary the heights slightly to give depth to the arrangement (Figure 16-17).

Step 8. Add flowers as needed to fill in the area between the center flowers and the side flowers. The placement of flowers should create a rounded form similar to the mound arrangement (Figure 16-18).

Step 9. Add greenery between the flowers to fill the spaces and cover the mechanics.

Step 10. Add small sprigs of baby's breath throughout the nosegay (Figure 16-19).

Step 11. Attach a bow with streamers to a 4-inch wooden pick. Insert the pick into the base of the holder (Figure 16-20).

Step 12. Check your work and evaluate the design.

FIGURE 16-16

Insert one mini-carnation between each of the first blossoms.

FIGURE 16-17

Place one to three flowers in the center to create a focal point.

FIGURE 16-18

Add additional flowers to create a mound form.

FIGURE 16-19

Add sprigs of baby's breath to the nosegay.

FIGURE 16-20
The complete nosegay.

Cascade Bouquets

The **cascade bouquet**, which is held at the waist, is created by extending the colonial bouquet into a flowing garland. The cascade may be any length from one foot to floorlength. The cascade bouquet is extremely popular with brides.

The cascade bouquet also can be made in several different ways. Large, heavy cascade bouquets may be hand-tied,

FIGURE 16-21
The cascade bouquet.

with each flower wired and taped separately. Small cascade bouquets are usually designed in foam bouquet holders. The flowers on one side of the nosegay are extended to form the cascade (Figure 16-21 and the color insert).

A third method consists of wiring and taping the long cascade which is then added to the foam bouquet holder. This is the method to be illustrated in this unit.

Step 1. Select materials:

> miniature carnations
> statice
> baby's breath
> leatherleaf fern
> foam bouquet holder
> 24- or 26-gauge wire

Step 2.   Wire and tape miniature carnations of various sizes, adding to each a sprig of baby's breath and leatherleaf. Use 24- or 26-gauge wire so the cascade will be light and flexible.

Step 3.   Begin the cascade by selecting the two smallest flowers. Place one to the left and about 2 inches higher than the other. Tape the two together (Figure 16-22).

Step 4.   Continue adding flowers left and right until the cascade reaches the desired length. If more than one flower or color is used, blend the two together. Increase the size of the flowers as the cascade becomes longer. The spacing between flowers should decrease as you reach the end of the cascade (Figure 16-23). Trim off excess wire as you construct the cascade. This prevents the cascade from becoming bulky and heavy.

Step 5.   Insert the cascade into the bottom of the bouquet holder. Insert the wired and taped portion of the

FIGURE 16-22
Beginning the cascade.

FIGURE 16-23
Completing the cascade.

FIGURE 16-24
Adding the cascade to the bouquet holder.

FIGURE 16-25

The completed cascade bouquet.

cascade into the foam until it comes out the other side and the cascade is in the desired position (Figure 16-24). Bend the top of the cascade over the rib of the plastic cage holder and back into the foam.

Step 6. Design the top of the bouquet in the same manner as described for a colonial bouquet. Carefully blend the cascade with the rest of the bouquet (Figure 16-25). The diameter of the top should be in proportion to the length and width of the cascade.

Step 7. Check your work and evaluate the design.

CORSAGES AND BOUTONNIERES

Corsages and boutonnieres are normally worn by the ushers, the best man, groomsmen, parents, grandparents, servers at the reception, other close family members, and anyone else who has a part in the wedding. This may include the organist, the soloist, and the preacher if he or she does not wear a robe.

Roses and cymbidium orchids are favorite flowers for the mothers and grandmothers, but almost any flower may be selected. Boutonnieres for the men are often selected from a flower used in the bride's or attendants' bouquets.

CEREMONIAL DECORATIONS

Many decorations are elaborate, but for the ceremony simple ones can be equally attractive. Since most weddings take place in a church, the florist must be familiar with the church's policy regarding flowers. Many churches have a written policy regarding the use of the church for weddings and receptions. The florist should obtain a copy of this policy ahead of time and plan floral decorations accordingly.

A variety of different decorations may be used for the wedding ceremony. The florist should plan decorations that enhance the beauty of the total area from the entryway to the altar area.

Flowers for the ceremony should be attractive and not overcrowded. Consider the distance from which the altar decorations will be viewed. Several small arrangements may be attractive when seen from the front of the church, but may be difficult to appreciate from the back of the room. Avoid the use of lavender, violet, or blue flowers in the decorations. These colors tend to recede into the background and will not be visible from the back of the church. They will not be seen in the wedding photographs either. A few large, bold arrangements may be more effective than several small arrangements.

The Entry and Vestibule

Floral decorations should begin at the entry to the church or other locations of the wedding. The guest book table or

FIGURE 16-26
The guest book table.

stand is usually located in the vestibule. An arrangement of flowers or a bud vase should decorate the table (Figure 16-26). Tall ficus trees or other live plants might be used in the vestibule. Garlands and live plants might be used to frame the entry into the sanctuary. Arrangements of cut flowers may be placed on the furniture in the vestibule to complement the wedding theme (Figure 16-27).

Aisles and Pews

The simplest form of pew decoration is a bow made from number 9 or number 40 ribbon. White ribbon is most commonly used but other colors are acceptable.

The pew bows are often accented with flowers and foliage that complement the altar decorations (Figure 16-28). The

FIGURE 16-27

An arrangement placed in the church vestibule.

arrangement may be constructed in a floral foam cage and attached to the pew with large rubber bands, chenille stems, or custom-designed pew markers. Pew markers may be placed on every pew or, often just as effective, every second or third pew. If the bride is on a tight budget, she may elect to decorate only the first one or two pews to identify the seating area for the family.

FIGURE 16-28
A pew marker.

Aisle candelabra are frequently used in candlelight ceremonies. The aisle candelabrum contains a single candle in a hurricane globe (Figure 16-29). The candles should not extend above the globe because of the danger of fire. Flowers may be added to the candelabrum by attaching a floral foam bouquet holder or a floral foam cage holder (see color insert). Cascading arrangements of flowers or foliage and ribbons are appropriate.

FIGURE 16-29

An aisle candelabrum.

The Altar

The altar decorations are the focal area of the wedding and should focus attention on the center of the altar and the bridal party. These decorations should be large enough to be seen by each wedding guest. They will be viewed from long distances so large flowers in white or bright colors are most effective.

A single arrangement is often used on the center of a large altar table. This arrangement is usually a large, triangular or fan-shaped design. Freestanding altar arrangements often complement altar flowers. These arrangements may be placed in standing wicker baskets or on brass stands. Arrangements may also be designed in large vases and elevated on adjustable brass stands.

Candelabra are popular altar decorations (Figure 16-30). They are made of wrought iron or steel with a polished brass finish and are available in a variety of shapes, including the seven-branch, the fan, the diagonal, the spiral, the heart, and the double ring.

Dripless candles are preferred over standard wax candles. These candles are made with a metal casing in the shape of a candle. Inside the casing, a spring pushes a thin, replaceable, wax candle up to the top as it burns. This eliminates problems with dripping wax.

FIGURE 16-30
Wedding candelabrum.

Candelabra may be decorated in a variety of ways. The simplest way is with a simple bow or a swag of foliage with a bow attached. Candelabra bowls may be purchased to fit onto the candelabra stand. Fresh flowers and greenery can be arranged in the bowls. Floral foam cage holders may also be attached to the candelabra to hold flower or foliage arrangements.

Foliage may also be used to decorate the altar. Live ferns and large, potted plants, such as the ficus, may be placed among the candelabra. Cut greens arranged in foam cage holders or vases are often used in the place of live plants. Emerald or jade palm may be arranged in palm buckets or Styrofoam. Oregon fern is frequently used to make foliage arrangements attached to the candelabra or on stands throughout the altar area (Figure 16-31 and the color insert).

FIGURE 16-31
Altar decorations.

FIGURE 16-32
A kneeling bench.

The kneeling bench is frequently used at weddings and may be a part of the altar decorations. Bows, greenery, and flowers may be added to the sides of the kneeling bench (Figure 16-32).

RECEPTION DECORATIONS

The reception decorations can easily continue the wedding theme. The same degree of formality used in the wedding should be used in the decorations for the reception.

Receptions in different areas of the nation vary greatly. A formal wedding reception in some areas would involve a sit-down dinner. In other areas, the formal reception would call for extravagant hors d'oeuvres. An informal reception would require fewer decorations, with hors d'oeuvres, or cake and punch, served in a garden or church social hall.

Regardless of the location, floral decorations add to the beauty of the reception area and make it more appealing. Flowers are used throughout the following areas of the reception.

Serving Tables

Serving or buffet tables may have a variety of decorations, depending upon the degree of formality and the theme of

FIGURE 16-33

Serving table.

the wedding. Figures 16-33, 16-34, 16-35, and the color insert illustrate several different themes for decorating the serving table. Pedestaled silver containers and candelabra are often featured on the serving tables. Mirrors and votive candles are popular items also. Arrangements of flowers, foliage, and fruit are frequently used (Figure 16-36). The fruit serves a dual purpose as both food and decoration.

FIGURE 16-34

Serving table.

FIGURE 16-35
Serving table.

FIGURE 16-36
An arrangement of fruit and foliages.

The Cake Table

The wedding cake is a tradition at weddings. The cake is sometimes featured on the serving table, but most often is displayed on a separate table. The cake table can be simply decorated with the cake as the main decoration (Figure 16-37). It may also be elaborately adorned with floral arrangements and backed with lattice and living plants (Figure 16-38). The cake itself may be decorated in a variety of ways. Flowers are popular alternatives to the traditional bride and groom. Floral decorations can also be added between the layers and around the base of the cake as in Figure 16-37.

FIGURE 16-37

The wedding cake table.

Punch and Champagne Tables

Punch and champagne tables may be simply decorated. Flowers and foliage in a circle at the base of the bowl are frequently used. If the table is large enough, floral arrangements may also be included. If a champagne fountain is used, the fountain usually has a container built into the top specifically to hold flowers.

Floating Pool Arrangements

A floating arrangement might be used as a decoration whenever a reception area features a pool. The design may feature a fountain with a low, round arrangement of flowers (Figure 16-39). Spotlights add to the beauty of the design if the reception is held at night.

Additional Reception Hall Decorations

Additional decorations may be placed throughout the reception hall. These may be as simple as the placement of

FIGURE 16-38

The wedding cake table with additional floral decorations.

FIGURE 16-39

A floating pool arrangement.

large potted plants or include elaborate arrangements of flowers, potted plants, columns and lattice (Figure 16-40). Arrangements can be placed throughout the room (Figure 16-41). They may even hang from the ceiling (Figure 16-42).

FIGURE 16-40

Reception hall decorations.

FIGURE 16-41
Reception hall
arrangement.

FIGURE 16-42
An arrangement woven
into a chandelier.

FIGURE 16-43

The rehearsal dinner banquet hall.

REHEARSAL DINNER

The rehearsal dinner is another part of the wedding calling for decorations. These may be as simple as a low arrangement for the head table or as elaborate as a completely decorated banquet hall (Figure 16-43).

Decorations for the tables should be kept low so as not to block the view of guests. Candles are often featured on the tables. One idea for table decorations is small floral arrangements placed in the china given to the couple as a wedding gift (Figure 16-44). Other decorations might feature live plants and lattice panels placed throughout the banquet hall.

DELIVERY OF THE WEDDING FLOWERS

Delivery of the wedding flowers is of utmost importance. A delivery schedule should be completed for each segment of the wedding: rehearsal dinner, wedding ceremony, and reception. The bride may specially request that the candelabra be placed in the church prior to the rehearsal. Special requests should be noted so that delivery can be completed on time.

Large items such as altar arrangements may be constructed on site. In this case, extra time must be allowed.

Many florists use a color coding system for their wedding work. Delivery boxes, floral stock, and rental equipment are

FIGURE 16-44
Guest table decorations.

color coded with colored stickers or waterproof markers. This allows for easy identification for the wedding and assures that items are not left behind. Items left at the shop mean return trips and delivery delays.

Student Activities

1. Construct each of the bouquets outlined in this unit. Try creative variations of each. For example, make the colonial nosegay in the form of an oval.

2. Invite a bridal consultant and a florist to talk to the class about current trends in wedding flowers and colors.

3. Ask a local florist to demonstrate the construction of a simple bridal bouquet and discuss how to work with a bride in planning her wedding floral decorations.

4. If your school has a home economics class that plans a mock wedding, volunteer to construct the flowers for the ceremony.

Self-Evaluation

A. Multiple Choice

1. A wedding order form is useful
   a. to plan floral decorations.
   b. as a means of organizing the wedding order.
   c. as a contract between the florist and the bride.
   d. all of the above.

2. _____ would be common flowers to include in a bridal bouquet for a bride with a traditional wedding gown.
   a. Birds-of-paradise
   b. Roses
   c. Tulips
   d. Proteas

3. The arm bouquet is also called a
   a. presentation bouquet.
   b. lazy man's bouquet.
   c. colonial bouquet.
   d. traditional bouquet.

4. Colonial nosegays are constructed in a _____ form.
   a. cascading
   b. natural
   c. circular
   d. crescent

5. Colonial nosegays may be constructed by using
   a. the hand-tying method.
   b. the foam bouquet holder.
   c. water picks.
   d. both a and b.

6. When constructing a cascading bouquet as outlined in this unit, construct the _____ first.
   a. top
   b. cascading portion
   c. bow
   d. focal point

7. _____ would be an example of a person who would require a corsage or boutonniere at a wedding.
   a. The mother of the bride
   b. The mother of the groom
   c. An usher
   d. all of the above

8. _____ usually tend to be more effective in the altar area.
   a. Arrangements with dark flowers used for accent
   b. Several small arrangements
   c. Foliage arrangements
   d. Large, bold arrangements

9. The guest book table or stand is usually located in the _____ of the church.
   a. sanctuary
   b. vestibule
   c. outside
   d. social hall

10. _____ decorations are the focal area of the wedding.
    a. Reception
    b. Vestibule
    c. Altar
    d. all of the above

11. _____ candles are the most preferred in a wedding.

    a. Dripless

    b. Handmade

    c. Cheap

    d. Wax

12. Arrangements composed of _____ can be used on serving tables.

    a. foliage

    b. fruit

    c. flowers

    d. all of the above

B. True or False

_____ 1. The florist should have a special area away from the traffic in the shop for the purpose of wedding consultations.

_____ 2. The wedding order form does not differentiate between flowers charged to the bride and groom because it is the bride's responsibility to pay for all of the flowers.

_____ 3. A bride should choose the type of bouquet she wants regardless of the style of her gown.

_____ 4. A large bride would want a small bouquet in order to make herself look slender.

_____ 5. Arm bouquets, colonial nosegays, and cascading bouquets are all popular bouquet styles.

_____ 6. Because of the way arm bouquets are constructed, they can be made in advance and preserved in vases of water or water picks.

_____ 7. A nosegay cannot be made well in advance of the wedding because it does not have a water source.

_____ 8. Cascading bouquets may be hand-tied.

_____ 9. Boutonnieres are usually made with flowers used in a bride's or bridal attendants' bouquets.

_____ 10. A florist should be familiar with a church's policy regarding weddings and receptions.

_____ 11. Dark flowers such as blue or lavender tend to create a lovely accent in arrangements when photographed.

_____ 12. Bows can make a simple pew decoration.

_____ 13. Because receptions are for the enjoyment of the wedding guests, the decorations should be less formal than the decorations for the wedding ceremony.

_____ 14. Avoid placing flowers on a wedding cake because you could poison the wedding guests.

C. Short Answer Questions

1. How can the florist determine how much a bride can afford to pay for wedding flowers?

2. How does the wedding order form assist in planning wedding decorations?

3. What criteria may be used in selecting the bridal bouquet?

4. Why should the florist avoid using several small arrangements in a large church?

5. How are the hand-tied bouquet and the arm bouquet similar and how are they different.

Sympathy Flowers

OBJECTIVE

To make sympathy floral designs.

Competencies to Be Developed

After completing this unit, you should be able to:

- identify the different types of sympathy flowers.
- construct a standing spray.
- construct a sympathy wreath.

Introduction

Sympathy flowers are another important segment of the retail florist business. Sympathy flowers include floral pieces that are sent to the funeral home as well as permanent flowers and potted plants placed on cemetery plots.

Small shops such as those located in shopping malls often do not provide sympathy flowers as part of their services. However, sympathy flowers have been referred to as the "bread and butter" income for many small flower shops. The reason is that such flowers are not requested seasonally but are needed throughout the year, providing a steady income for the shop.

Competition among flower shops for sympathy flower business can be keen. For this reason, the designer should use the same degree of talent and creativity when arranging sympathy pieces as with any other specialty design.

A new shop or new owner-manager must learn the customs and traditions of the local area in relationship to sympathy flowers. These may vary greatly from region to region. The florist must also cooperate with both customer and funeral home director to satisfy their preferences.

Flowers are sent as a tribute to the deceased and as an expression of love and sympathy for the living members of the family and friends. Since the time between a death and the funeral is usually only two to three days, orders must be taken, designed, and delivered quickly. This is in addition to other work that the shop receives so it can mean extra hours of work at night and on weekends.

Some shops locate near large funeral homes and specialize in sympathy flowers. They are usually located in larger cities where the volume of sympathy flower business is larger.

Close family members will meet with the florist to plan the major floral arrangements, mainly the **casket cover**, the floral piece that sits on the casket lid. This is a very emotional time for the family. Many do not have specific ideas about the types of flowers desired or the style of the floral tribute. Often, they are dependent on the florist's information and suggestions about sympathy flower designs.

BASIC SYMPATHY DESIGNS

A variety of floral designs are used as sympathy flowers and the designer must be able to make each creatively. The size and style of the design will depend on the request of the customer, the types of flowers used, and the flower budget. Sympathy flowers often provided for funerals include casket covers, standing sprays, wreaths, baskets, arrangements, and decorated potted plants.

Each funeral piece should be tagged with a card giving the name of the deceased and the sender. Double cards are best. Half of the card remains on the floral piece and half can be removed and given to the family of the deceased. A description of the floral piece should be included on the card to help family members remember the piece when sending acknowledgments. The address of the sender is also helpful.

FIGURE 17-1

A three-quarter couch casket cover.

Casket Covers

The casket cover is the floral piece that sits on the lid of the casket couch (Figure 17-1 and the color insert). This floral piece is purchased by the closest family members. It is also the most elaborate and beautiful of all the sympathy flowers.

Three sizes of casket covers are available: the full couch, three-quarter couch and the half-couch. Most common are the three-quarter and half-couch sizes. Full couch sprays are expensive and difficult to handle during the funeral service if the casket is to be opened.

Sprays

Sprays are the most popular funeral pieces ordered (Figure 17-2). These may be single or double. The single spray has a bow at one end with flowers extending to the other end. They are designed to be displayed on a special frame that holds up to one dozen sprays along the sides or end of the casket. Single sprays are less expensive than double sprays and wreaths, offering the customer a broader price range from which to select. Figure 17-3 shows the placement of the main flowers to be placed in a single spray. The price of the spray determines the types of flowers used. A bow is added and filler flowers placed around the bow and between the primary flowers.

FIGURE 17-2
A double standing spray.

FIGURE 17-3
Placement of the main flowers in a single spray.

Double sprays are called standing sprays because they are displayed on wire easels. **Dixon picks** are used to attach the Styrofoam to the easel. There are two wooden picks attached on opposite ends of a flexible metal strip. Double sprays are often placed alongside the casket, and the flowers may be coordinated to match the casket cover. The spray is usually an oval, diamond, or triangular shape. Most often a bow is placed in the center with flowers extending in both directions.

The techniques for constructing sprays are variable and each shop will have its own unique style. Sprays may be hand-tied with the natural stems showing. Sprays constructed in this manner do not have a source of water so they must be delivered the day of the funeral. In many regions of the country, family members receive guests the night before the funeral. Family and friends usually want their flowers displayed for this occasion. This makes the use of hand-tied arrangements difficult.

A second method of constructing sprays calls for the use of floral foam. This procedure has the added benefit of staying fresh longer. The foam can be secured in a holder, such as the Oasis Floracage holder and the John Henry spray bar (Figure 17-4). Both of these companies manufacture floral products. These holders can be easily attached to an easel for construction and display.

A third method of constructing sprays is with the use of a block of Styrofoam. Many florists prefer this method be-

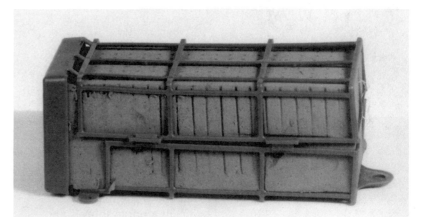

FIGURE 17-4

A floral foam holder used in the construction of sprays.

FIGURE 17-5

Flowers placed in water picks.

cause of the strength of the Styrofoam. Styrofoam holds the flowers much better than floral foam. This added security is important when the sympathy pieces are being transported to the funeral home, from the funeral home to the church service, and from the church to the grave site.

Flowers to be used in sprays may be placed on a wooden or steel pick for insertion in the Styrofoam. This procedure can only be used when delivered the day of the funeral. More common is the placement of flowers in water picks (Figure 17-5). **Water picks** are small plastic tubes with a small reservoir of water and a rubber cap. The cap is slit so that flowers can be inserted into the reservoir of water. The picks have a pointed end that can be pushed into the Styrofoam spray bar.

The size of the spray and the numbers and kinds of flowers used depend upon the price of the spray. A basic double spray may be constructed in the following manner.

Constructing a Standing Spray

Step 1.   Select materials:

Styrofoam spray bar (4″ × 12″)
wire easel
2 wooden picks
dixon picks
water picks
14 carnations
1 bunch of pompons
2 to 3 bunches of Emerald or Jade palm
2 bunches of plumosa

FIGURE 17-6
Press dixon picks into the back of the Styrofoam where the frame of the easel crosses the Styrofoam.

FIGURE 17-7
Green the back of the spray bar with palm.

Step 2.   Secure the Styrofoam block to a tripod easel by pushing the top section of the block onto the hook of the easel. Press dixon picks into the back of the Styrofoam where the frame of the easel crosses the Styrofoam (Figure 17-6). Press wooden picks into the base of the Styrofoam and tie the wires around the easel frame.

Step 3.   The palm used to green the back of the spray can be attached to steel picks. A slanted cut at the end of the palm will also allow it to be placed directly into the Styrofoam (Figure 17-7). Shorten the stems of the palm at the sides so the spray is elongated.

Step 4.   The carnations will be added to the spray in the pattern shown in Figure 17-8. Cut and wire the carnations to the lengths shown, placing each in a water pick.

Step 5.   Add the carnations to the spray in the pattern illustrated in Figure 17-9 by pressing the water pick into the Styrofoam until it feels secure.

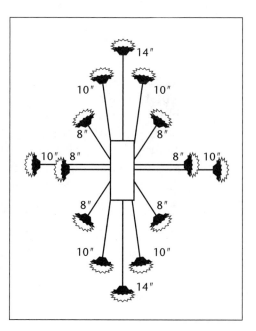

FIGURE 17-8

The pattern and lengths of flowers for a double spray.

FIGURE 17-9
Add carnations to the spray.

FIGURE 17-10
Complete the greening of the spray.

Step 6. Cut the stems of palm in half and place the top half over the bottom half. Add a sprig of plumosa on top of the palm and secure it with a steel pick. Many florists will attach the plumosa to picks and add it separately. Either method is acceptable. Complete the greening of the spray (Figure 17-10).

Step 7. Cut pompons about 8 inches in length. Place them two at a time into water picks (Figure 17-11).

Step 8. Tie a large loose bow, attach it to a wooden pick, and add it to the center of the spray.

Step 9. Add the pompon picks to the spray beginning near the bow and working outward in the spaces be-

FIGURE 17-11
Place pompons into water picks.

FIGURE 17-12
Add the pompon picks to the spray.

tween the carnations (Figure 17-12). Place some flowers and greenery among the loops of the bow.

Step 10. If desired, filler flowers such as baby's breath can be added to soften the appearance of the spray.

Wreaths

Sympathy wreaths are constructed in a similar manner to sprays except the piece is circular in design (Figure 17-13 and the color insert). The wreath form may be made of Styrofoam, straw, or a plastic wreath casing filled with floral foam. If the straw wreath is used, it must be wrapped with green plastic wreath wrap. All of these may be purchased

FIGURE 17-13
A sympathy wreath.

from wholesale florists in sizes ranging from 12 to 30 inches. The wreath size used will depend upon the price specified by the customer.

Constructing a Wreath

Step 1. Select materials:

a 14-inch or 16-inch Styrofoam or straw wreath form
21 carnations
2 bunches of pompons
3 bunches of Emerald or Jade palm
2 bunches of plumosa

Step 2. Mount the wreath onto the easel. Press the top of the wreath onto the easel hook. Wire the wreath to the top of the frame of the easel. Repeat this where the bottom of the wreath crosses the frame of the easel.

Step 3. Cut sprigs of palm in half and attach the two halves together using a steel pick. Place these onto the wreath in a circular pattern (Figure 17-14).

Step 4. Add a piece of plumosa to additional sprigs of palm and complete the greening of the wreath (Figure 17-15). The palm and plumosa can be placed on the wreath separately.

Step 5. Tie a large, loose bow and attach it to a wooden pick. The bow can be placed at the bottom center of the wreath, the top center, or to the left of center.

FIGURE 17-14
Green the outside of the wreath.

FIGURE 17-15
Complete the greening of the wreath.

FIGURE 17-16
Add carnations to the wreath.

FIGURE 17-17
Add pompons to the wreath.

Step 6. Cut the stems of the carnations to about 8 inches in length. Wire each carnation and insert the stem into a water pick.

Step 7. Add the carnations, beginning at the bow and working around the circle (Figure 17-16). Check to see that the water picks are completely covered in the greenery. Add additional sprigs of plumosa if needed.

Step 8. Cut the pompons to 8-inch lengths and place them in water picks as directed in the section on sprays. Add the pompon picks to the wreath. Begin at the bow and work around the circle (Figure 17-17).

FIGURE 17-18

A variation of the wreath is achieved by adding gladioli around the bow.

FIGURE 17-19

A wreath with a Bible.

Step 9. The wreath may be used as illustrated or filler flowers added, such as baby's breath. A variation could be achieved by adding a few gladioli wired to wooden picks around the bow (Figure 17-18).

Many variations of the sympathy wreath are possible. A designer must be creative with funeral designs so that each piece going to the same funeral will look different. Figure 17-19 shows a wreath with an open center containing a Bible. The Bible can be removed after the funeral and kept as a memento of the deceased. Part of the wreath ring can be covered with ribbon or galax leaves and the flowers placed in a crescent or S-shape (Figure 17-20).

FIGURE 17-20

A crescent-shaped wreath.

Baskets

The funeral basket is a traditional sympathy tribute in most areas of the country (Figure 17-21 and the color insert). Baskets may be made out of plastic, metal, or papier-mâché. Grapevine baskets with handles are sometimes used by placing a plastic liner inside the basket. After the flowers have died, the basket can be used to decorate the home.

The most commonly used flowers for funeral baskets are gladioli, carnations, mums, and pompons. Sometimes roses and lilies will be added to increase the value of the basket. The pattern of the basket arrangement is usually triangular or fan-shaped, but other creative designs can be used. Bows are often added to the arrangement, but should not be the dominant feature. We will not construct a funeral basket arrangement in this unit because the triangular and fan-shaped arrangements have been covered in a previous unit.

FIGURE 17-21
A funeral basket.

Other Floral Designs

The retail flower shop can expect to receive orders for many kinds of sympathy pieces. If the deceased belonged to a fraternal organization or lodge, that group will often request their insignia be used at the service. Styrofoam forms representing the most common fraternal organizations can be purchased from a wholesale florist. Examples of these include the Masonic Lodge and the Order of the Eastern Star.

Hearts and crosses are also popular in sympathy arrangements (see color insert). The Christian cross design is one of the more traditional funeral pieces (Figure 17-22). It is often

FIGURE 17-22

The Christian cross design is a very traditional funeral piece.

backed with gathers of satin ribbon and the face of the cross is covered with tightly placed pompons. Other flowers, such as carnations and roses, are added in a diagonal line or S-shape across the cross.

Living plants are also frequently ordered as an expression of sympathy. In the spring, pots of tulips and azaleas are popular as are poinsettias at Christmas. Various foliage plants are also decorated and used at funeral services. Many people like to send living plants so that the family can use them later to decorate their homes. The decoration of potted plants is described in another unit.

Student Activities

1. Constructing funeral pieces is difficult in a floral design class because you have no use for the piece afterward. Try constructing a wreath using foliages you can gather from your yard or the woods. Try using magnolia or loquat leaves. Small sprigs of laurel or oak leaves could be used to green the arrangement. Use silk flowers instead of live flowers so that they can be used a number of times. Be creative with the materials at hand.

2. Invite a local florist to the class to demonstrate making sympathy pieces. You might have to wait until the florist has funeral orders but not such a large number that he or she does not have time to come to the class.

3. Invite a local funeral director to talk to the class about the role of flowers at funeral services and the funeral home's policy regarding the handling of flowers.

Self-Evaluation

A. Multiple Choice

1. _____ are considered the "bread and butter" income of many small florists.
   a. Holidays
   b. Sympathy flowers
   c. Weddings
   d. none of the above

2. The most elaborate of all sympathy flowers, which are also purchased by the closest family members, is the
   a. casket cover.
   b. casket couch.
   c. standing spray.
   d. funeral wreath.

3. _____ are the most popular funeral pieces ordered.
   a. Casket covers
   b. Funeral wreaths
   c. Sprays
   d. Baskets

4. Double sprays are also called
   a. oval sprays.
   b. triangular sprays.
   c. double trouble.
   d. standing sprays.

5. Sprays may be constructed using
   a. hand-tying methods.
   b. floral foam.
   c. Styrofoam.
   d. all of the above.

6. Sympathy wreaths are constructed in a manner similar to sprays except
   a. wreaths are more expensive.
   b. wreaths require more labor.
   c. wreaths are circular in design.
   d. none of the above.

7. _____ is a material commonly used to make baskets.
   a. Metal
   b. Straw
   c. Papier-mâché
   d. both a and c

8. A _____ is a flower commonly used in funeral baskets.
   a. carnation
   b. orchid
   c. tulip
   d. gerbera daisy

9. The pattern of a basket arrangement is usually
   a. triangular.
   b. fan-shaped.
   c. crescent-shaped.
   d. both a and b.

10. Many people like to send living plants in sympathy because
    a. they are cheaper.
    b. they can be used later by the family to decorate their home.
    c. they are much easier for the florist to decorate.
    d. they symbolize the beauty of life.

B. Short Answer Questions

1. Make a list of the different kinds of sympathy flowers.
2. Why have sympathy flowers been referred to as the "bread and butter" of the flower shop business?
3. How do sympathy flower orders affect the work day of flower shop employees?
4. What is the advantage of placing flowers in water picks when designing sympathy flowers?

Drying Flowers

desiccant-drying
glycerin
silica gel

bleach
borax
fine sand
glycerin
kitty litter
microwave oven
preservative spray for
  dried flowers
Rit dye
silica gel
white cornmeal

OBJECTIVE

To dry and preserve flowers and foliage to make a dried flower arrangement.

Competencies to Be Developed

After completing this unit, you should be able to:

- select and dry flowers by the hanging method.
- select and dry flowers by the desiccant-drying method.
- preserve flowers and foliage using the glycerin method.
- bleach and dye dried flowers and foliage.

Introduction

Dried and preserved flowers and foliages are often sold in the flower shop to customers desiring a permanent arrangement (Figure 18-1). Several methods of preservation have been developed, and almost any flower or foliage can be preserved. As a result, florists have a wide variety of materials available to them.

Even though most flowers and foliages can be preserved, few florists elect to dry and preserve their own materials. They prefer to buy them from commercial sources.

Some retail florists may choose to preserve floral products themselves. Drying and preserving flowers is also an

FIGURE 18-1
Dried flower
arrangement.

excellent project for a floral design class. At the end of your flower arranging class, you should have a large variety of flowers dried and preserved to make a beautiful flower arrangement. By drying your own flowers, you are not restricted to using only what is blooming at the time.

The most common methods of drying and preserving flowers will be presented in this unit.

HANGING METHOD

Many flowers dry well by this simple process. Pick flowers just before they reach their prime. Overly mature flowers do not dry well. Strip off all leaves and tie the flowers in small

FIGURE 18-2

Hanging method of drying flowers.

bunches. Suspend the flowers upside down in a warm, dry area without light (Figure 18-2). Darkness preserves the color. An attic or basement that is warm and dry is an ideal place.

Drying time depends on the type of flower being preserved and the atmosphere of the place where they are drying. Generally, most flowers dry in 1 to 3 weeks using the hanging method. Leave bunches hanging until they are needed for arrangements. The following flowers and foliages dry well using the hanging method.

astilbe	gypsophila	rabbit tobacco
celosia	larkspur	roses
delphinium	liatris	Scotch broom
dock	milkweed	statice
goldenrod	millet	strawflower
grasses	pussywillows	yarrow

DESICCANT-DRYING

Desiccant-drying consists of burying flowers in a substance that will extract moisture from the flowers by absorption.

Almost any flower can be dried without loss of color or shape when the petals are supported by one of a number of drying agents. The function of the support medium is to allow even drying throughout the flower and to keep petals from curling. Therefore, all spaces between the petals must be completely filled.

To dry flowers by this method, first remove the stems from the flowers about one-half inch below the calyx. Place one to two inches of the drying agent in the bottom of a container. Place the flowers face up on the bed of desiccant. Then, cover the flowers completely with more of the desiccant by gently supporting the petals and working the substance into and between the folds of the flowers.

Seal the container to prevent the drying agent from absorbing moisture from the atmosphere. Strong cardboard boxes make excellent containers because holes can be punched into the bottom of the box to remove the drying agent without damaging the flowers.

The length of time required for a flower to dry varies. Normally 2 to 4 weeks are required for complete drying. Experimentation is suggested for determining the drying time for specific flowers using different desiccants.

When flowers have dried completely, gently remove them from the container and brush them with a small paint brush to remove particles and dust from the petals. Flowers dried by this method are extremely fragile. Spray them with a dried flower preservative to protect and strengthen them.

Attach floral wires to the flowers for stems or the original stems can be dried by hanging and then reattached to the flower with hot glue. If wires are attached for stems, use floral tape to secure the flower to the wire and to give the stem a more natural look. The following is a list of flowers that can be dried by desiccant-drying.

anemone	dahlia	marigold
aster	daisy	pansy
butterfly weed	day lily	pompons
calendula	dusty miller	rose
camellia	gazania	tithonia
carnation	hollyhock	verbena
cosmos	larkspur	zinnia

Sand and Borax

A mix of sand and borax may be used as a drying agent. Use fine, washed beach sand for this purpose. It may be purchased from building supply stores. The sand should be sifted well before use. If the sand is damp, it can be oven-dried in a shallow pan at 250° F for thirty minutes. Measure out two parts sand and one part borax and mix well. Borax can be purchased in the laundry section of any grocery store. Drying time is one to two weeks for this material.

Cornmeal and Borax

White cornmeal and laundry-grade borax also make an excellent drying agent. This mixture is light and works well with delicate flowers. Mix ten parts of white cornmeal with three parts borax. Sift to mix thoroughly. The borax is added to protect the petals from mold and weevils during the drying process. This method takes approximately three to seven days.

Kitty Litter

Kitty litter made of ground clay may be used as a drying agent. The ground clay has a great deal of absorbing quality and it can be used over and over. Sift or screen the kitty litter and discard larger pieces. Select a brand of kitty litter that contains small particles. Some brands are very coarse and unsuitable for drying flowers.

Silica Gel

Silica gel is an industrial compound that may be purchased for drying flowers. It is more expensive than other drying agents, but it can be reused almost indefinitely. Silica gel is probably the best drying agent for preserving flowers because it dries quickly and the flowers retain more of their natural colors. It may be purchased at hobby stores or discount stores which carry crafts.

The silica gel sold for drying flowers has blue crystals, known by the brand name Tell-Tale, added as a color indicator. When the crystals turn pink, it indicates that the silica gel has absorbed the maximum amount of moisture. It may be sifted to remove flower debris and dried in an oven at 250° F for thirty minutes. The Tell-Tale crystals will return to their blue color when completely dried.

MICROWAVE DRYING

The microwave oven enables you to dry flowers in a matter of minutes. The quality is superior. The colors are brighter and the flowers are not as dry and perishable as conventionally dried flowers. Use the following steps as a guide for drying flowers of your own.

Step 1. Select materials:

> brightly colored flowers, partially open
> silica gel or uncolored kitty litter
> 18-gauge wire
> floral tape

Step 2. Trim stems 1/2- to 3/4-inch in length. Spread silica gel 1 to 2 inches deep in small glass or paper bowls. Do not use metal in the microwave. Prepare one container for each flower (Figure 18-3). This makes removal of the flowers easier.

FIGURE 18-3

Spread silica gel 1 to 2 inches deep in a small glass bowl.

FIGURE 18-4

Sprinkle silica gel between petals until flowers are completely covered.

Step 3. Arrange one flower blossom, face up, in each bowl of silica gel. Sprinkle additional silica gel between petals until the flowers are completely covered. Use a toothpick or small brush to separate petals if needed (Figure 18-4).

Step 4. Place one or two flowers in the microwave oven at a time. Place a cup of water in the corner of the oven to provide moisture in the oven. This will prevent the flower from completely drying up. Microwave for 1 to 4 minutes according to the flower drying guide (Figure 18-5). These are approximate

FIGURE 18-5

Flower drying guide for a microwave oven.

FLOWER DRYING GUIDE			
Flower	No. of Flowers	Power Level	Total Time (Minutes)
Carnation	2	High	3 to 3-1/2
Daisy	2	High	1 to 2
Pompons	2	High	3 to 4
Rose	1	High	2-1/2 to 3
Camellia	1	High	3 to 4
Marigold (large)	1	High	3 to 4

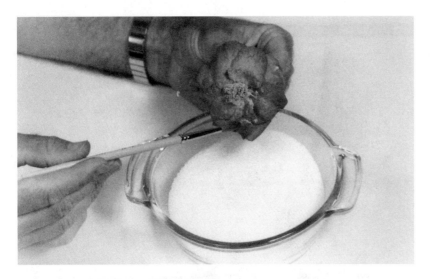

FIGURE 18-6

Use a small paintbrush to remove silica gel from flower.

times. Some experimentation will be required to determine exact times. After heating, let the flowers remain in the medium for 1 to 24 hours.

Step 5.   Remove flowers carefully from the bowl. Shake gently to remove silica gel. Use a small paintbrush to remove any remaining gel (Figure 18-6). Spray with a preservative spray and attach a 16- or 18-gauge wire to the stem with floral tape.

GLYCERIN

Glycerin is a colorless liquid made from fats and oils that can be used to preserve foliage. The advantage of using glycerin is that it preserves the foliage in a pliable, more natural state. It is frequently used to preserve such greenery as eucalyptus, magnolia, and maple leaves. Glycerin can also be used to preserve some flowers such as baby's breath. Glycerin may be purchased from any drug store.

Make the solution by mixing one part glycerin and two parts water. Pour this into a container to a depth of 4 to 5 inches. The stems of fresh foliage should be given a fresh slanting cut at the base. Place the stems of the foliages in the glycerin solution for 4 days to 2 weeks (Figure 18-7). Replenish the solution as needed during the preserving process. The foliage will darken to an olive or bronze color as the

FIGURE 18-7

Glycerin method of preserving foliages.

leaves absorb the solution. To prevent this darkening of the leaves, add absorption dyes to the water. The dye will be absorbed into the stem and deposited in the foliage. When the process is complete, hang the foliages upside down to dry.

Individual leaves are best preserved by submerging the leaves in the glycerin solution. Place weights on the leaves to keep them below the surface. Ivy leaves and stems may be preserved by this method. Leave the ivy submerged for 4 days, and it will turn a lovely shade of green. As soon as the foliage is removed from the solution, rinse the glycerin off with cool water.

BLEACHING AND DYEING

Materials to be bleached must first be dried. Place the plant material in a solution of 1 cup of bleach to 2 gallons of water. Use plastic, glass, or enamel containers. Do not use metal containers. Weight material down so it will be submerged at all times. Leave 5 to 6 days or longer if necessary. Check periodically. Remove when bleaching has occurred, and rinse thoroughly. Leave the material in a water bath for a day or two. Hang outside to dry and whiten (Figure 18-8).

To dye bleached materials, dip for about five minutes in a boiling Rit solution according to the directions on the package. Rinse the material after dyeing. Hang to dry out of the sun.

FIGURE 18-8

Wooden stems whitened by bleaching.

Materials suitable for bleaching include many of those that can be air-dried. Aspidisdra and bird-of-paradise leaves are suggested foliages. Grape and wisteria vines bleach well but must first be peeled. Okra and corn shucks are other items often bleached.

Student Activities

1. Collect a variety of native weeds and grasses to dry by the hanging method.
2. Desiccant-dry the same kind of flower, using several different drying agents, and compare the results.
3. Collect and preserve several kinds of foliage using the glycerin method.
4. Dry and preserve a variety of flowers and foliage. Save these to be used in a later unit on arranging permanent flowers.

Self-Evaluation

A. Multiple Choice

1. When drying flowers using the hanging method, it is best to suspend the flowers in a dark area because
   a. darkness speeds up the drying process.
   b. darkness helps preserve the color of the flowers.
   c. darkness stops photosynthetic activity in the flowers.
   d. darkness repels insects that might damage the flowers.
2. Most flowers dry in _____ when using the hanging method.
   a. 1 to 3 weeks
   b. 1 to 3 days
   c. 1 to 3 months
   d. 6 to 9 months
3. _____ is an example of an agent used in desiccant-drying.
   a. Kitty litter
   b. Glycerin

    c. Silica gel

    d. a and c

4. Drying flowers using _____ requires the least amount of time.

    a. a desiccant only

    b. a desiccant and a microwave

    c. the hanging method

    d. glycerin

5. When using sand or cornmeal as a desiccant, add _____ to the mixture.

    a. silica gel

    b. bleach

    c. borax

    d. kitty litter

6. In order to preserve foliage in a more natural, pliable state, place the stems in _____

    a. borax.

    b. bleach.

    c. antifreeze.

    d. glycerin.

7. _____ is/are an example of an item that is desirable to bleach.

    a. Corn shucks

    b. Grape vines

    c. Okra

    d. all of the above

B. True or False

_____ 1. Most florists choose to dry flowers on their own.

_____ 2. When drying by the hanging method, it is best to choose flowers that have not yet reached their prime.

_____ 3. It is best to remove stems from flowers before drying them with a desiccant.

_____ 4. Any kind of kitty litter makes a suitable desiccant for drying flowers.

_____ 5. When Tell-Tale crystals in silica gel are pink, it is time to dry the gel out in the oven.

_____ 6. Avoid putting water in the microwave when drying flowers because the desiccant tends to absorb the extra moisture.

_____ 7. Use a small paintbrush to remove dust and desiccant granules after desiccant-drying.

_____ 8. Glycerin will turn many foliages a dark bronze color.

_____ 9. Bleach floral materials before attempting to dry them.

_____ 10. Bleached materials can be dyed using food coloring.

C. Short Answer Questions

1. List four materials that can be used for desiccant-drying.
2. Describe the environment where flowers should be hung for drying.
3. What are the advantages of microwave drying?
4. Describe the procedure for glycerin treatment of flowers.

Unit **19**

Arranging Permanent Flowers

OBJECTIVE

To construct arrangements using permanent flowers.

Competencies to Be Developed

After completing this unit, you should be able to:

- identify the different types of permanent flowers.
- state two methods of making silk flowers.
- identify paper flowers.
- identify a number of dried and preserved flowers.
- make a permanent flower arrangement.

Introduction

A wide variety of permanent flowers are available to the retail florist. The quality and variety of permanent flowers is continually improving and they are frequently requested by customers.

Permanent flowers include silk, paper, and dried or preserved flowers. The design principles that guide the construction of fresh flower arrangements also apply to permanent flowers. However, the mechanics involved in arranging permanent flowers are different. Permanent flowers are easier to arrange because they are not perishable and may be manipulated and mechanically secured without concern for wilting.

Terms to Know

freeze-dried flowers
paper flowers
silk flowers

Materials List

Examples of molded, plastic-stemmed silk flowers and hand-wrapped silks
Examples of paper flowers
Examples of different kinds of dried flowers

SILK FLOWERS

The term **silk flowers** applies to a wide variety of fabric flowers that are sold to florists. Silk, nylon, cotton, rayon, and blends of each are commonly used. There is also a great variability in the grades of fabric utilized to create silk flowers, thus, a variety of quality and price levels.

Silk flowers are available in two basic types: molded, plastic-stemmed flowers and hand-wrapped silks. Molded flowers are usually made of polyester, which is heat molded in a die form to create the petals. These are placed on a stem made of wire covered with plastic. Molded flowers are usually less expensive than hand-wrapped silks. These silks are usually produced with multiple flowers on a single stem. Silk bushes and bouquets are frequently made with this type of silk flower (Figure 19-1).

FIGURE 19-1

Molded, plastic-stemmed silk flowers.

The petals of hand-wrapped silks are usually cut with a stamp, then attached to wire and positioned to create the flower. The flower is then secured to a wire stem that is hand-wrapped with floral tape. The flowers may be hand-painted or hand-dyed, a process that makes the flowers look real. Hand-wrapped silks are more expensive because of the labor required to produce them. These flowers are usually sold individually (Figure 19-2).

FIGURE 19-2
A hand-wrapped silk flower.

FIGURE 19-3
A paper flower.

PAPER FLOWERS

Paper flowers are created from rice paper, parchment, and bark fiber paper, which have a high fiber content (Figure 19-3). They are formed in the flower shape and most are available with wired petals and leaves. The flowers and leaves are attached to a main wire stem, then wrapped with paper or floral tape. Paper flowers may be predyed before being formed into flowers or airbrushed after the flower has been formed. Paper flowers are also more expensive than molded, plastic-stemmed flowers because of the amount of labor required to produce them.

DRIED AND PRESERVED FLOWERS

Dried and preserved flowers are in great demand by customers for permanent arrangements in the home. Dried flowers are frequently used for wall hangings and for decorating wreaths (Figure 19-4). Dried materials can also be combined with silk flowers to create attractive permanent arrangements (Figure 19-5).

Most of the methods for drying and preserving flowers have been described. One additional method is discussed here.

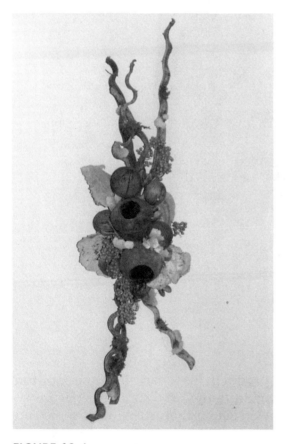

FIGURE 19-4

Dried flowers are frequently used for wall hangings.

FIGURE 19-5

Dried materials can be combined with silk flowers to create permanent arrangements.

FIGURE 19-6
A freeze-dried flower.

FREEZE-DRIED FLOWERS

Freeze-dried floral products have had all of the moisture mechanically removed from their cell structures (Figure 19-6). However, the flowers retain their shape with some suppleness to their texture. Freeze-dried floral products are available from most wholesalers. The equipment for freeze-drying is expensive, so the cost of these flowers is higher than that of flowers dried by other methods.

IDENTIFYING DRIED AND PRESERVED FLOWERS

Most flowers and foliage can be easily dried or preserved. However, most florists prefer to purchase dried materials from a commercial source. Purchasing dried and preserved flowers and foliage can be confusing because they are sometimes packaged without labels, and distributors sometimes give different names to the same product in different areas of the country. The pictures in Appendix C will help you to identify some of the most commonly used preserved materials.

MECHANICS FOR PERMANENT FLOWERS

The mechanics used when arranging permanent flowers are similar to those for fresh flowers. However, precautions against flowers wilting are unnecessary. A dry foam made especially for permanent flowers is used to support the stems. Two types of dry foam for permanent flowers are available. One is used for delicately stemmed flowers and the other for larger, sturdy stems. Styrofoam may also be used. Styrofoam is strong and will hold heavy or large stems but is difficult to use with delicate stems. Delicately stemmed flowers are usually placed on a wooden or steel pick before being inserted into the Styrofoam.

Dry foam can be wedged into containers without additional support. In containers where wedging the dry foam is impossible, the foam may be attached to the container with a floral adhesive, such as hot glue. Anchor pins glued to the container may also be used to secure it.

The dry foam or Styrofoam will need to be hidden by covering it with sheet moss or Spanish moss. This will prevent the foam from being visible between the flowers. Use greening pins to hold the moss in place. Flower stems can be easily inserted through the moss.

The stems of silk or dried flowers may be glued in place as the design is constructed. This will prevent the stems of heavy flowers from turning and from slipping out of the foam. Dipping stem ends in pan glue before inserting them is probably the easiest method.

FIGURE 19-7
Cover the foam with moss and attach with greening pins.

FIGURE 19-8
Establish the skeleton of a scalene triangle using three of the smaller flowers in the silk bushes.

Constructing A Basket Arrangement of Permanent Flowers

Step 1. Select materials:

a fireside basket
2 silk bushes of your color choice
1 silk bush of filler flowers
dry foam (1/3 of a block)

Step 2. Prepare the container. Cut one-third of a block of dry foam. Place the foam in the bottom of the basket with a wooden pick or small stick across the top. Run a wire through the basket and over the foam. Tie the wire securely to hold the foam in place. Cover the foam with sheet moss or Spanish moss and attach with greening pins (Figure 19-7).

Step 3. Establish the skeleton of a scalene triangle arrangement using three of the smaller flowers in the silk bushes (Figure 19-8).

FIGURE 19-9
Add mass flowers.

FIGURE 19-10
Add filler flowers.

Step 4. Add mass flowers following the pattern of the scalene triangle. Do not crowd the flowers (Figure 19-9).

Step 5. Add small filler flowers in the open spaces between the mass flowers (Figure 19-10). Add additional greenery, if needed, to give the arrangement depth and cover mechanics.

Step 6. Check your work and evaluate the arrangement using the rating scale in Appendix G.

Constructing a Contemporary Arrangement

Use this activity as a guide only. Do not try to make your arrangement look exactly like the one illustrated. Select similarly sized materials which reflect your own taste in flowers and color. See Figures 19-11 and 19-12 for additional examples of permanent flowers.

FIGURE 19-11

Permanent arrangement using silk and dried flowers.

FIGURE 19-12

Permanent arrangement using silk and dried flowers.

Step 1. Select materials:

> 3 large, hand-wrapped magnolia blossoms or other large flowers
> 12 dried poppy seed heads
> 1 twig of curly willow
> 1 stem of hand-wrapped raspberries or other small fruit or flower
> silk foliage
> container
> dry foam

Step 2. Prepare the container. Select a tall container. Cut one-third of a block of dry foam and trim it to fit the container opening. Wedge or glue the foam to secure it in place. Cover the dry foam with sheet moss or Spanish moss and use greening pins to hold it in place.

Step 3. Insert a sprig of curly willow several times the height of the container into the center back of the foam.

Step 4. Add the magnolia blossoms. Cut the first blossom to about three times the height of the container, and place it just to the left of center. Leave this blossom about half open. Place the second blossom in the center just above the rim of the container. This flower should be fully open. Position the third flower downward and to the right of center. Curl the flower head forward, and open the flower about three-fourths of the way (Figure 19-13).

Step 5. Add the dried poppy seed heads and silk raspberries in a staggered pattern around the magnolia blossoms as shown in Figure 19-14.

Step 6. Place silk greenery around the base of the arrangement, including the back (Figure 19-15).

FIGURE 19-13
Add magnolia blossoms.

FIGURE 19-14
Add dried poppy seed heads and silk raspberries in a staggered pattern.

FIGURE 19-15
Place silk greenery around the base of the arrangement.

Step 7. Check your work and evaluate the arrangement using the rating scale in Appendix G.

Student Activities

1. Visit a flower shop and examine the arrangements of permanent flowers.
2. Sponsor an identification contest in the class over dried and preserved materials. Have a small permanent arrangement as the prize for the winner.
3. Complete the two activities outlined in this unit and then design an arrangement using your own creative ideas.

Self-Evaluation

A. True or False

_____ 1. The term *silk flowers* applies to a wide variety of fabric flowers made from silk, nylon, cotton, and rayon.

_____ 2. Silk flowers are produced by two basic methods: hand-molded stems and plastic-wrapped stems.

_____ 3. Permanent flowers are more difficult to arrange than fresh flowers.

_____ 4. Molded, plastic-stemmed flowers are more expensive than hand-wrapped silk flowers.

_____ 5. Dried flowers that have had all of the moisture mechanically removed from their cell structures are called freeze-dried flowers.

_____ 6. The same floral foam used for fresh flowers is recommended for permanent flowers.

_____ 7. The foam used to hold permanent flowers is usually covered with moss held in place with anchor pins.

_____ 8. The ends of permanent flower stems may be dipped in pan glue to prevent them from moving in the dry floral foam.

B. Short Answer Questions

1. Describe the two types of silk flowers.
2. What materials are used to hide the dry foam or Styrofoam when arranging permanent flowers?
3. What are freeze-dried flowers?

Selecting Indoor Plants

Terms to Know

binomial name
common name
drenching
scientific name
taxa
taxon

Materials List

*a variety of foliage
and flowering
potted plants.*

OBJECTIVE

To assist customers in selecting appropriate indoor plants for various occasions.

Competencies to Be Developed

After completing this unit, you should be able to:

- identify the most common indoor flowering and foliage plants.
- distinguish between temporary and permanent indoor plants.
- match plants' light requirements with the proper environment.

Introduction

Indoor potted plants are an important segment of the items for sale in the retail flower shop. Many people prefer potted plants to cut flowers because they last longer. Many customers who send flowers choose potted plants so that the recipient may enjoy the blooms for a long period of time.

As an employee in a flower shop, you must be able to care for the plants in the shop. You will also need to be knowledgeable about potted plants in order to help your customers make appropriate selections. Customers want to be educated about the plants they select and may ask a multitude of questions. They often want to know the name of the plant, if it's a flowering or foliage plant, its light and water requirements, and its life expectancy. If you can answer these questions, your chances of making a sale are much greater.

PLANT NAMES

Indoor potted plants are identified by a common and a scientific name. The name given to a plant by the people living in an area is called the **common name.** The common name often refers to some use made of the plant or some unusual characteristics. For example, the common name of one plant is watermelon peperomia because the leaves of the plant resemble the markings of a watermelon. Common names may become confusing because the same plant may be called by different common names in different areas of the country. For this reason, potted plants are often bought and sold by their **scientific name.**

The scientific name of a plant comes from a classification system based on how plants are related to each other. The system was developed by a Swedish botanist named Carolus Linnaeus in 1743. The same basic system is still used today.

All living things are divided into two kingdoms, the plant kingdom and the animal kingdom. The plant kingdom is divided into about a dozen major divisions. Each division is divided into classes and each class into subclasses and orders. The breakdown continues through family, genus, species and varieties. The name given to a plant in each category is always in Latin or Greek with the exception of the variety name. In Figure 20-1, find the scientific classification of the tulip.

Each of the plant categories is called a **taxon.** A group of plant categories is called a **taxa.** As you can see from the classification of the tulip, the taxa are divided into major and minor taxa. Horticulturists are most concerned with the minor taxa.

One of the contributions Linnaeus made to plant classification was a simple naming system using the genus taxon and the species taxon. This system is called the **binomial**

	CATEGORY (TAXA)	CLASSIFICATION
	Kingdom	Plant
	Division	Spermatophyte
Major Taxa	Class	Angiosperm
	Subclass	Dicotyledon
	Order	Lilale
	Family	Liliaceae
	Genus	Tulipa
Minor Taxa	Species	Fosteriana
	Variety	Red Emperor

FIGURE 20-1

Scientific classification system.

system, meaning that it uses two names to identify plants. There are international rules which regulate the way plants are named. They stipulate that when a binomial name is given to a plant, it cannot be used for any other species of plant.

As an employee of a flower shop, you will want to learn the names of plants. With customers, you will probably utilize the common name. If you are ordering plants for the shop, you may need to use the scientific name of the plant to avoid confusion relating to common names. In this book, both names will be given to identify plants.

LIGHT REQUIREMENTS

Most customers of a flower shop buy a plant because of its appearance and price. They give little thought to the requirements of the plant. It is essential, however, that the plant be right for the light it will have in its new home. As a salesperson, you need to make customers aware of this information. This may result in a more satisfied customer. If the plant performs poorly because of improper lighting, the customer will be unhappy and may even blame the flower shop. If the plant does well, then the customer may buy additional plants in the future.

The following information will assist you in understanding the light requirements of plants.

Type of Light	*Description*
High light	This includes plants that grow best in full sun or bright, indirect light, such as that found in or near sunlit windows and in places where there is strong, reflected light.
Medium light	Plants in this group grow best in a bright, but sunless, window, or four to eight feet from a sunny window.
Low light	Plants in this group grow well with indirect light, such as that in a shaded window, or at a point more than 8 feet from a bright window.

WATERING

More plants die from overwatering than any other cause. For this reason, you need to be able to explain to customers how to water the plants they purchase.

Each plant has individual watering needs. The proper frequency of watering is not constant. It depends on the size of the plant, the size of the container, the environment, and the time of year. Scratch the top one-half inch of the soil surface to determine if a plant needs water (Figure 20-2).

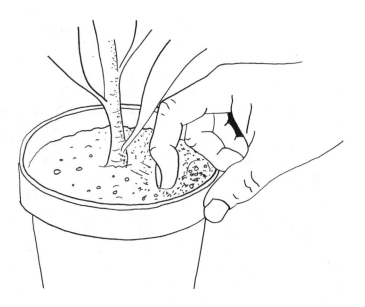

FIGURE 20-2

Check plants often to determine their water needs.

In this unit, the moisture requirements of plants are divided into the following groups.

Type of Plant	*Description*
Dry-in-winter plants	Desert cacti and succulents should be treated as moist/dry plants during the active growing season from spring to autumn. During the winter, the soil should be allowed to dry out almost completely between waterings.
Moist/dry plants	The recommendation for this group is to water thoroughly and frequently between spring and autumn, and to water sparingly in winter, letting the top 1/2 inch of soil dry out between waterings.
Moist-but-not-wet plants	Most flowering plants belong in this group. The soil is kept moist, but not wet, at all times. Water carefully each time the surface dries, but never frequently enough to keep the soil saturated.
Wet-at-all-times plants	Very few plants belong in this group. Water thoroughly and frequently enough to keep the soil wet, not merely moist. Examples of plants in this group are azaleas and umbrella plants.

The best way to determine when to water is also the simplest. Look at the surface of the soil, weekly in the winter and daily in the summer if possible. If the surface is dry and powdery all over and the plant should be kept moist at all times, water it. For other plants, insert your finger about 1/2 inch into the soil. If the soil is dry at this depth, then the plants need watering. The most important exceptions are the cacti and succulents in winter. If the room is cool, leave them alone unless there are signs of shriveling.

When watering plants, water until liquid runs out the bottom of the pot. This is called watering to **drenching**. This way you can be assured the entire volume of soil is wet. If the pots are sitting in drip trays, empty the trays after 30 minutes.

PERMANENT OR TEMPORARY PLANTS

If a cyclamen starts to die after a couple of months, it's not because the customer has done anything wrong. The cyclamen is a temporary plant, and the salesperson needs to explain that at the time of purchase.

Many flowering potted plants fall into this category (Figure 20-3). These plants are purchased to be enjoyed for a short time and then be discarded. Some can be made to bloom again the next season, but this requires a skilled gardener.

An important group of flowering potted plants known as "gift" plants fall into this category. The azalea, gloxinia, cyclamen, and chrysanthemum fall into this group. Other temporary potted plants are the garden bulbs, such as daffodils and tulips. With practically all of the plants in this group, the flowers will fade after a few weeks, then the leaves will fall. It is a basic feature of these plants.

Other flowering potted plants, such as the African violet and peace lily, bloom continuously or cycle through blooming periods during the year (Figure 20-4). Many of these, such as the peace lily and the anthurium, are attractive foliage plants when not in bloom. Flowering potted plants in this book will be divided into temporary and permanent groups.

FIGURE 20-3

A temporary flowering potted plant.

FIGURE 20-4

A permanent flowering potted plant.

IDENTIFYING POTTED PLANTS

Appendices D, E, and F are provided to help you identify the more common potted plants and to learn basic information about them. The more you know about the plants, the better you will be able to assist your customers in making appropriate selections. These appendices are intended to help you get started. Examine live specimens whenever possible.

Student Activities

1. Play games to help the class learn the plants in this unit and make learning more fun. Divide the class into teams and provide each team with a bell or whistle. Show flash cards of the plants or live specimens and see which team can most quickly name the plant and the category to which the plant belongs. Keep track of the points and provide small prizes to the winning team.

2. Visit a wholesale florist or commercial greenhouse to observe the many plants available in your area.

3. Participate in FFA Plant Identification Contests. If none are available in your area, start one at your school.

Self-Evaluation

A. Select the best answer from the choices offered to complete each statement.

1. Questions commonly asked by customers purchasing plants include
   a. the name of the plant.
   b. how much light a plant needs.
   c. how often the plant should be watered.
   d. whether it grows in the greenhouse.
   e. all of the above.
   f. a, b, and c.

2. The common names for plants can be confusing because
   a. they are difficult to pronounce.
   b. more than one plant may have the same common name.
   c. they are similar to the scientific name.
   d. growers do not know plants by common names.

3. The scientific name for plants is called the
   a. trinomial name.
   b. plant order.
   c. binomial name.
   d. subclass.

4. The binomial name for a plant includes the
   a. class and subclass.
   b. order and family.
   c. division and class.
   d. genus and species.

5. The scientific name of a plant is written in
   a. English.
   b. Greek or Latin.
   c. Spanish.
   d. English or Latin.

6. A plant that grows best in bright indirect light should be placed
   a. near a north window.
   b. in a south window.
   c. in an east window.
   d. in a west window.

7. Watering to drenching means watering until
   a. the entire plant is wet.
   b. water drips off the leaves.
   c. the top of the soil is covered with water.
   d. water runs out the bottom of the pot.

8. A gloxinia is classified as a
   a. temporary flowering potted plant.
   b. permanent flowering potted plant.
   c. temporary foliage potted plant.
   d. permanent foliage potted plant.

B. Match each of the plants in the list below with the correct plant classification.

_____ 1. cyclamen	a. temporary flowering potted plant
_____ 2. African violet	
_____ 3. peace lily	b. permanent flowering potted plant
_____ 4. tulips	
_____ 5. lollipop plant	
_____ 6. Easter lily	
_____ 7. gloxinia	
_____ 8. kalanchoe	
_____ 9. poinsettia	
_____ 10. hibiscus	

C. Short Answer Questions

1. Why is it important for the retail salesperson to be knowledgeable about plants sold in the shop?

2. How can you best determine if a plant should be watered?

Decorating Potted Plants

Terms to Know

florist foil
jardiniere
plant face
pot-et-fleur
speed covers

Materials List

clearphane
   transparent film
dish garden
florist foil
number 3 and number
   9 ribbon
pompon
   chrysanthemums
preformed pot covers
a variety of potted
   plants
water picks
wooden picks

OBJECTIVE

To decorate potted plants.

Competencies to Be Developed

After completing of this unit, you should be able to:

- decorate a potted plant using florist foil.
- decorate potted plants using preformed pot covers.
- decorate potted plants using a jardiniere.
- decorate dish gardens.

Introduction

You do not have to read this unit to know about the beauty and popularity of house plants. Just look around you. You will find them in office buildings, shopping malls, and probably your own home. Green plants bring the freshness and natural beauty of the out-of-doors into interior locations.

Because of their natural beauty, potted plants are important items for sale in the retail flower shop. Many different kinds of plants are sold including architectural, or standing floor plants, climbing plants, table top plants, hanging plants, and dish gardens. Figure 21-1 gives the names of some of the most popular plants in each of these categories.

Architectural Plants

Dieffenbachia

Dracaena

Fatsia

Norfolk Island Pine

Palms

Pony Tail Plant

Rubber Plant

Tree Philodendron

Weeping Fig

Yucca

Hanging Plants

Christmas Cactus

Creeping Fig

Ferns

Goldfish Plant

Grape Ivy

Piggyback Plant

Spider Plant

Velvet Plant

Climbing Plants

Arrowhead Vine

Heart Leaf Philodendron

Pothos

Split Leaf Philodendron

Stephanotis

Dish Gardens

Aluminum Plant

Fittonia

Golden Pothos

Heart Leaf Philodendron

Ivy

Peperomia

Pilea

Prayer Plant

Sanseveria

Table Top Plants

Achimenes

African Violets

Anthurium

Azalea

Bromeliads

Chinese Evergreen

Cyclamen

Gloxinia

Hydrangea

Kalanchoe

Peace Lily

Peperomia

Rex Begonia

FIGURE 21-1

Types of plants sold in the retail flower shop.

Many people purchasing plants for the home or office will choose potted plants rather than cut flowers because of the lasting beauty of the plants. This is also true for many people selecting a gift for a business open house, the hospital, or as a sympathy flower. The plant is a long-lived gift that can be later used to decorate the home or office. This is an excellent selling point for the florist to make. Potted plants that are already decorated make a quick and easy sale during busy hours.

Potted plants purchased by retail flower shops are usually grown in 4-, 5-, or 6-inch pots. Larger foliage plants will necessarily be in larger pots. The pots are usually made of plastic and sometimes clay. Sometimes the plants are sold just as they are received from the grower as cash-and-carry plants. Most often the pots will be decorated to make them more attractive. Several types of materials may be used to decorate potted plants. A number of these will be presented in this unit.

POLY FOIL

One of the most popular materials used by florists to cover pots is called **poly foil**, or pot wrap. It is a plastic-coated aluminum foil that may be purchased in rolls 20 inches wide and 30 feet long. Poly foil comes in a variety of colors, to harmonize with the flower and foliage color, and it may be purchased in various thicknesses. The heavier foils are less likely to tear but cost more. These are often used with larger potted plants.

Poly foil is available in plain or textured patterns. Regardless of the pattern or color, most foil is green on the backside to more closely match the color of the plant foliage.

Wrapping Pots Using Poly Foil

Step 1. Water the plant and allow it to drain. Remove any broken leaves and spent flowers. Clean the foliage of the plant with warm water if needed. Limit the use of leaf shine products as these may be harmful to the plant.

Step 2. For taller plants, cut a piece of foil so that it extends up two-thirds the height of the plant (Figure 21-2). The foil on shorter, bushy plants should extend just over the top of the pot.

Step 3. Fold over all edges of the foil 1/2 inch for a more finished appearance.

Step 4. Choose the "**face**," the most attractive side, of the plant and position the pot in the center of the foil with the face toward you.

FIGURE 21-2
Florist foil should extend two-thirds the height of taller plants.

Step 5. Lift the foil in front of the pot with your left hand and lift the corner of the foil with your right hand. Pull the corner toward the center of the pot tucking excess foil between the pot and the corner of the foil. Form a crease well back from the corner of the foil (Figure 21-3).

Step 6. Repeat step 5 with the left side of the foil, overlapping one side. Insert a wooden pick through the overlapping layers of foil to hold them in place (Figure 21-4). The two overlapping sides can also be stapled or held into place with hot glue.

Step 7. Turn the plant around, and repeat steps 5 and 6 for the opposite side.

FIGURE 21-3
Crease the foil.

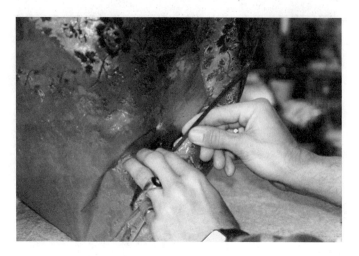

FIGURE 21-4

Insert a wooden pick through overlapping layers of foil.

Step 8. Gathers may be made in the sides if the foil is excessively loose.

Step 9. Cut a 30-inch piece of number 9 ribbon in a color that harmonizes with the color of the flower or foliage. Tie a double knot over the wooden pick at the front of the pot (Figure 21-5).

Step 10. Tie a bow and attach it to a wooden pick.

Step 11. Insert the wooden pick, with bow attached, under the loop of ribbon and into the pot (Figure 21-6). The original pick may be removed or left in place. Position the bow to complete the wrapping (Figure 21-7).

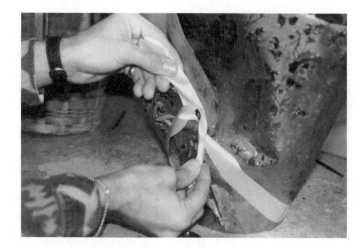

FIGURE 21-5

Tie a double knot over the wooden pick.

FIGURE 21-6

Insert the wooden pick and bow.

FIGURE 21-7

Pot decorated with poly foil.

PREFORMED POT COVERS

Preformed pot covers may be purchased in a variety of materials from aluminum foil to straw. They are more expensive than poly foil but save on labor because they are so easy to use. **Speed covers** are preformed aluminum foil pot covers. They are available in a variety of colors and sizes. The potted plant is simply placed into the speed cover, and a bow is added.

Accessory materials such as clearphane transparent film and Mylar film can be added to give the potted plant a more finished appearance (Figure 21-8). Clearphane film is available in a variety of sizes, colors, and patterns. Popular patterns include the French lace and polka dot motifs, or designs.

Mylar film is also available in a variety of colors and sizes. It may be purchased in rolls or in 18-by-30-inch sheets. Mylar film is made from the same materials as Mylar balloons and is more expensive than clearphane film. When used as an accessory to a potted plant, it gives the finished pot a flashy contemporary look.

Decorating plants with speed covers and clearphane film or Mylar film is quick and easy. For a 5- or 6-inch plant pot, cut a square of film from a 20-inch roll. Place the pot in the middle of the film. Gather the film up by lifting the four corners and gently drop the pot into a speed cover (Figure 21-9).

FIGURE 21-8

Clearphane transparent film and Mylar film.

FIGURE 21-9

Lift the four corners of the film and drop the pot into the speed cover.

FIGURE 21-10

Add the bow.

Add a bow as previously instructed, and the task is complete (Figure 21-10).

If the top of the soil is visible in a potted plant, add sheet moss or Spanish moss to cover the soil. If the plant is short and bushy, it may be more desirable to simply insert the bow into the pot from the top rather than the side (Figure 21-11).

Straw plant covers with plastic liners, often called straw hats, may also be used to decorate plants. These are also available in a number of sizes including those that will cover large potted plants. The procedure for decorating plants in a straw plant cover is essentially the same as that for speed covers (Figure 21-12).

FIGURE 21-11

For short bushy plants, insert the bow into the pot from the top.

FIGURE 21-12

Decorated plant with straw plant cover.

Many florists have learned to be creative in decorating potted plants. Their ideas distinguish florists from each other. Figure 21-13 shows how one florist creatively decorated a pot of tulips using a straw pot cover. Stems of curly willow have been attached to the sides of the pot cover using wire and raffia. The two stems of willow have been tied together across the top of the pot. Eucalyptus, grapes, and lotus pods have been added to further decorate the pot. What was a simple pot of tulips has now become an elaborate arrangement of flowers and dried materials. Creative decorating such as this can be used to upgrade the value of any potted plant.

FIGURE 21-13

Pot of tulips decorated with straw plant cover.

JARDINIERES

A **jardiniere** (jar-duh-neer´) is a decorative container used to hold potted plants. Jardinieres are made from a number of materials including pottery, china, and plastic. Also popular are jardinieres made from brass.

Jardinieres greatly increase the cost of a potted plant. However, they save money in the long run, as they can be used again with future cash-and-carry plant purchases. This is a good selling point to the customer.

Jardinieres do not have a drainage hole in the bottom so caution must be used in watering plants. Place gravel, perlite, or a plastic liner in the bottom to keep the plant from sitting in drainage water.

Jardinieres should be proportionate in size to the pot, with the ideal pot size just slightly smaller than the jardiniere. If the jardiniere is too large, the plant will look

FIGURE 21-14
Flowering plant displayed in jardiniere.

FIGURE 21-15
Foliage plant displayed in jardiniere.

out-of-scale. Figures 21-14 and 21-15 show examples of plants displayed in jardinieres.

DECORATING DISH GARDENS

Dish gardens are shallow containers which hold a number of living plants crowded together (Figures 21-16 and 21-17). Plants are selected which have the same light requirements so that they will live and grow together. Dish gardens can easily be made from plants rooted and grown in 2- to 3-inch pots. These are crowded together in a container and additional soil is added to fill the empty spaces. Some florists prepare their own dish gardens but most purchase them from wholesalers.

Dish garden containers do not have a hole for drainage in the bottom so use extreme caution when watering them.

FIGURE 21-16
Dish garden.

FIGURE 21-17
Dish garden.

If water stands in the bottom of the container, the plant roots will die.

The European dish garden is a variation of the traditional dish garden. When making a European dish garden, the plants are left in the pot, wedged into a container and held in place with Styrofoam or moss (Figure 21-18). Small flowering plants are often used along with foliage plants to add

FIGURE 21-18
Constructing a European dish garden.

FIGURE 21-19

A small flowering potted plant adds a splash of color to a European dish garden.

a splash of color to the dish garden. Small African violets, primulas, and kalanchoes are excellent additions to a dish garden (Figure 21-19).

Cut flowers can be added to a dish garden for additional color. When these temporary additions of color are made, the resulting combination is called a **pot-et-fleur.** When making a European dish garden, a small piece of wet floral foam can be added among the plants. Cut flowers can then be placed in the foam. They may also be added by using water picks (Figure 21-20). Fill the water picks with water and add stems of flowers. The water picks can then be pressed

FIGURE 21-20

Cut flowers may be added to dish gardens with water picks.

FIGURE 21-21

Water picks filled with water and flowers are pressed into the soil of the dish garden.

into the soil among the foliage plants (Figure 21-21). Add as many water picks as desired and finish the pot-et-fleur with the addition of a bow (Figure 21-22).

A dish garden is always a welcome present to receive, but it is not a permanent indoor garden. If the dish garden contains flowering plants, wait until the flowers fade, then break up the arrangement of plants and repot the individual plants before the roots become hopelessly intertwined.

FIGURE 21-22

Pot-et-fleur with the addition of a bow.

CARE OF POTTED PLANTS IN THE SHOP

Potted plants are not usually kept in the flower shop for a long time, but they can decline quickly if not given proper care. Employees must know how to ensure that plants remain healthy and attractive.

When plants arrive at the shop, they should be groomed by wiping the leaves with a damp cloth to remove residues. Also, dead blooms and damaged foliage should be removed before the plant is decorated for display.

Light

Adequate light can be a problem in maintaining healthy plants in the shop. Care should be given to display plants in areas where they receive some natural light from shop windows. Supplemental light can be provided by the use of lamps. Try to group plants by their light requirements and display them in areas of the shop which best meet those requirements.

Watering

Watering can be a challenge in a flower shop where large numbers of plants are often displayed. Plants displayed in decorative containers are often overwatered because drainage water collects in the container. This causes the roots to rot quickly and the plant leaves to turn yellow and drop off or wilt. Plants that do not receive enough water turn light green or pale yellow. Both underwatering and overwatering can cause plants to look undesirable, making them unmarketable. This can mean a loss for the shop.

Feel the pot's soil often to see if the plants need water. See Appendices D, E, and F for the water requirements of specific plants. If plants feel dry, water until water runs out the bottom of the pot into the pot liner. Do not let the plant sit in drainage water.

Plants that are wrapped in poly foil or dish gardens without drainage holes are difficult to water properly. These plants should not be waterlogged, but rather, given enough water to moisten all the soil. This may take some experimenting to learn just how much water to apply.

Temperatures

Warm temperatures inside the shop can cause flowering plants to mature and fade quickly. Flowering plants last longer at cooler temperatures than those found in flower shops. Though it is difficult, try to avoid placing flowers near air conditioning vents where they will be exposed to warm, dry air. Also, avoid putting plants in areas exposed to drafts from cold air and to extreme changes in temperature as might be found near a window.

Student Activities

1. Practice wrapping potted plants with florist foil using the steps outlined in this unit as your guide. Because measuring foil is a little different for larger plants than it is for smaller bushier plants, be sure to wrap various sizes of plants as you practice.

2. Decorate pots using a variety of preformed pot covers. Discuss the advantages and disadvantages of using each type of pot cover.

3. Have a florist demonstrate to the class unique techniques for arranging and decorating dish gardens. Each student can then create his or her own dish garden. Optional: The class can have a dish garden contest, with students competing in various categories such as "most unique," "best design," and so forth.

Self-Evaluation

A. Select the best answer from the choices offered to complete the statement or answer the question.

1. Florist foil
   a. comes in a variety of colors.
   b. is available in plain or textured patterns.
   c. is almost always green on the back side to match plant foliage.
   d. all of the above.

2. A jardiniere is
   a. rarely used because of its cost.
   b. a decorative container used to hold flowers.
   c. a type of bow popularly used with dish gardens.
   d. a container used for European dish gardens.

3. Plants used in dish gardens
   a. must all be the same color.
   b. must be foliage plants.
   c. need to be watered frequently.
   d. must all have the same light requirement.

4. Preformed pot covers are popular with florists because
   a. they are inexpensive.
   b. they take up relatively little storage space.
   c. they are timesaving and easy-to-use.
   d. they look good without bows.

5. When adding cut flowers to a dish garden, you would need to use _____ to prolong the life of the flowers.
   a. wet floral foam
   b. water picks
   c. sphagnum moss
   d. both a and b

6. The use of flowers as a temporary addition to a dish garden results in a combination called a
   a. pot-et-fleur.
   b. dish garden.
   c. jardiniere.
   d. European dish garden.

7. When covering taller plants with florist foil, cut the foil so that it extends _____ up the height of the plant.
   a. 3/4   c. 1/2
   b. 2/3   d. 1/3

8. The most attractive side of the plant is also called the _____ of the plant.
   a. head   c. face
   b. view   d. front

9. If the soil is visible after covering a pot, use _____ to cover the soil.
   a. clearphane   c. Spanish moss
   b. sheet moss   d. both b and c

10. A bow
    a. must always be made with a wide ribbon.
    b. can be inserted in a plant from the top or the side.
    c. is an expensive decoration for a plant.
    d. must be made of cloth.

B. Short Answer Questions

1. How would you convince a customer to purchase a plant in a jardiniere even though it costs more?
2. What is a pot-et-fleur?
3. What is the advantage of using speed covers instead of foil?
4. Name two types of preformed pot covers.
5. What is a key selling point for potted plants versus cut flowers?

Pricing Strategies

OBJECTIVE

To calculate the cost of floral products.

Competencies to Be Developed

After completing this unit, you should be able to:

- calculate the cost of goods for floral items.
- use three pricing strategies to calculate the retail value of floral items.
- define leader pricing.

Introduction

The objective of any retail flower shop is to make a profit for the owner of the shop. Since the retail florist faces heavy competition, determining the right price can be difficult. If the owner-manager sets prices too high, then the flower shop may lose customers to other shops. If prices are set too low, the flower shop may fail to make a profit.

Many factors contribute to making pricing a difficult job. The variability in the price of flowers during the year is one large factor. Flowers are reasonable when in season, but become very expensive out of season, at which time wholesale prices may double or even triple.

Terms to Know

divisional percentage pricing
leader pricing
nesting pricing
ratio markup plus labor
standard ratio markup pricing

Materials List

Wholesale price list for a variety of flowers and floral products

Holiday demand for flowers also makes pricing difficult. The wholesale price for flowers greatly increases during holiday periods such as Valentine's Day, Mother's Day, and Easter.

A shop that sets a standard price for a dozen roses will find a great difference in profit at various times of the year. The pricing strategy for a shop must take all of these factors into account if the shop is going to make a profit.

DETERMINING THE UNIT COST OF GOODS

If a flower shop is to successfully use any method of pricing, it is necessary to determine the cost of every item used. The following examples are only a guide for calculating the individual cost of goods. You should use your own cost-of-goods figures to calculate similar unit costs.

Almost all hard goods, such as ribbon or containers, are sold in cases, boxes, rolls, or other bulk amounts. A florist normally uses these products by the piece, foot, or some other unit of measure. The total cost of these goods needs to be broken down into single unit cost. This is easily done by dividing the total cost by the number of units enclosed. Always round up any fraction of a number. If you do not round up, the shop will lose money.

Example 1:   number 3 corsage ribbon
Cost ÷ number of yards per roll = price per yard
$4.00 ÷ 100 yards = $.04 per yard

Example 2:   floral container
Cost ÷ number of pieces = price/piece
$22.50 ÷ 24 = $.9375 or $.94/piece

The same formula can be used to calculate the cost per stem when working with fresh flowers.

Example 3:   fresh gladioli
Cost ÷ number of stems = price/stem
$5.95/bunch ÷ 10 stems = $.595 or $.60/stem

Example 4:   leatherleaf fern
Cost ÷ number of stems = price/stem
$1.69 ÷ 25 = $.0676 or $.07/stem

Some flowers such as spray chrysanthemums should be priced by the flower instead of the stem because the number of flowers on a stem is so variable. Standardized figures for the number of flowers per bunch may be obtained from your wholesaler or you may make a count of the flowers in a bunch.

Example 5:   Pompon chrysanthemums
Cost/bunch ÷ number of flowers = price per flower
$2.95 ÷ 30 blooms = $.098 or $.10/bloom

Some goods at a wholesale florist will be sold as a collection of items at a single price. For example, three containers that fit into one another might sell for a single price. The price of each item in the collection must be determined. The method used to do this is called the **nesting pricing method.** The name comes from the items being packed or nested one inside the other.

Example 6:   A nest of three containers of varying sizes costing $7.50

Step 1.   Number each item starting with 2 and add up the numbers.

Smallest container   2
Medium container   3
Largest container   4
Total number   9

Step 2.   Divide the purchase price by the total number. Always round up to the next number.

$7.50 ÷ 9 = $.833 or $.84

Step 3.   Multiply the resulting value by the number assigned to each container.

$.84 × 2 = $1.68 value of small container
$.84 × 3 = $2.52 value of medium container
$.84 × 4 = $3.36 value of large container

STANDARD RATIO MARKUP PRICING

The **standard ratio markup pricing** method has long been used by retail flower shops. This is also the easiest method of pricing. It is a flexible system which operates by determining the wholesale cost of an item and then multiplying, or marking up, that cost by a number from two to four or more to cover operating cost, labor, and profit. This system has worked because of years of experimentation in determining the multiplying factor, or ratio. However, it has a weakness: The florist can only make an educated guess at the appropriate ratio to use.

The ratio markup pricing system has been used with different ratios for different products. Nonperishable items such as giftware and supplies may be marked up at a 2-to-1 ratio. Perishable items and items requiring a lot of special attention may be marked up at a 4-to-1 or even 5-to-1 ratio. Below are some categories of markup ratios.

RATIO	TYPES OF ITEMS
2-to-1	Supplies such as containers, ribbons, floral foam, gift items, cash-and-carry cut flowers and plants.
3-to-1	Basic arrangements, bud vases, decorated plants, basic wreaths and sprays.
4-to-1	Creative arrangements, wedding flowers, corsages, reception and party flowers.

Items requiring a great deal of time and attention might even be marked up at a 5-to-1 ratio. The owner-manager or head designer must be able to determine what items belong in each category and then train the other salespeople or designers to do the same for this system to work.

Try this example. Using the sample wholesale price list given in Figure 22-1, calculate the wholesale cost of materi-

ITEM	WHOLESALE PRICE
Floral foam	$32.00/case (48 blocks)
Container	$18.00/dozen
Carnations	$6.25/bunch (25)
Pompons	$2.95/bunch (30 blossoms)
Statice	$3.95/bunch (10 stems)
Baby's breath	$5.95/bunch (12 stem/bunch)
Roses	$21.60/bunch (25)
Leatherleaf	$1.69/bunch (25 stems)
Anchor tape	$2.65/roll (180 feet)
Cardetts	$2.75/hundred
Enclosure Cards	$.75/fifty

FIGURE 22-1

Sample wholesale price list.

als, determine the appropriate markup ratio and retail price of an arrangement using the following materials.

ITEM	UNIT PRICE	WHOLESALE COST
1/3 of a block of floral foam	$ .67	$ .23
container	1.50	1.50
9 carnations	.25	2.25
15 blooms of pompons	.10	1.50
3 stems of statice	.40	1.20
12 stems of leatherleaf	.07	.84
1 foot of anchor tape	.02	.02
1 cardett	.03	.03
1 enclosure card	.02	.02
Total wholesale cost of materials		$7.59

This arrangement would be considered a basic arrangement so a 3-to-1 ratio would be used for markup. This would make the arrangement sell for $22.77 ($7.59 × 3). Most florists round the cost of arrangements to the next fifty cent or dollar figure. This would make the arrangement sell for $23.00.

The ratio markup pricing system can also be used to determine which goods can be used in arrangements of a certain price. For example, a customer orders a $35.00 nosegay for the school prom. This would be considered a creative design so the markup ratio would be 4-to-1. The cost of goods for this design would be $8.75 ($35.00 ÷ 4). This amount guides the florist's choice of flowers, ribbons, and other items to be used in designing the nosegay.

RETAIL COST OF GOODS PLUS LABOR

The **retail cost of goods plus labor** method involves calculating the retail value of each part of an arrangement using the ratio markup method and then adding a percentage for labor.

The use of a standardized labor percentage makes it easy for prices to be calculated. A range of percentages may be used depending on the amount of labor or expertise required to make the design. A higher percentage might be applied to a wedding bouquet than to a bud vase. A figure of 30 percent labor is commonly used by florists. The amount could be varied up or down depending on the design.

The following example illustrates the calculations of the retail price for a basic arrangement using the retail cost of goods plus labor method. The expertise to make a basic arrangement would not be very high so a 25 percent labor figure will be used in this example.

ITEMS	COST OF GOODS	× MARKUP	= RETAIL PRICE
container	$.80	× 2	= $1.60
flowers and foliage	5.25	× 3	= 15.75
cardett, enclosure card, tape flower care information tag	.10	× 2	= .20
			$17.55

$17.55 × 25 percent labor = $4.39 labor
$17.55 + $4.39 = $21.94, or $22.00.

DIVISIONAL PERCENTAGE PRICING METHOD

The **divisional percentage pricing** system includes net profit as a factor in pricing. Every floral shop has three controllable areas of expenses. They are operating expenses, labor, and cost of goods. The total of these, plus net profit, should equal 100 percent of a shop's gross sales. The following formula illustrates this principle with suggested percentages for each category.

Gross Sales	=	Operating Expenses	+	Labor	+	Cost of Goods	+	Net Profit
100%	=	35%	+	20%	+	30%	+	15%

The divisional percentage pricing system is based on the cost-of-goods percentage for the flower shop. The cost of goods may vary from shop to shop and the owner-manager should calculate that percentage for a particular shop.

This percentage is used to calculate the retail price of floral products by dividing the actual cost of goods by the cost-of-goods percentage. The following example illustrates this pricing strategy. The total cost of goods for an arrangement is calculated to be $8.25.

$8.25 ÷ 30 percent (cost-of-goods percentage)
= $27.50 retail price.

The retail price for this arrangement would be $27.50. Remember that the percentages a florist uses are individually determined. A change in the percentage will vary the retail selling price of an item.

LEADER PRICING

Leader pricing is a strategy used to help lure customers into the flower shop. With this strategy, certain items may be offered at prices well below their normal price. When customers visit the shop to purchase the leader item, they may purchase other items as well. Perhaps they may also become repeat customers for the shop.

SUMMARY

Three strategies for pricing floral items have been presented in this unit. This information should help you understand the importance that pricing strategies play in the operation of a flower shop.

Regardless of the type of pricing strategy selected by a shop, it is important that prices be clearly marked on all products. This may be done with signs or stick-on and tie-on types of tags.

Student Activities

1. Construct a wholesale price list for a number of floral items similar to the sample wholesale price list given in Figure 22-1. Use the current wholesale prices for your area.

2. Determine the unit cost for each item in the sample wholesale price list and for the list you constructed in the above activity.

3. Invite a local florist to share his pricing strategies with the class.

Self-Evaluation

A. Complete each of the following problems.

1. Use the ratio markup method to determine the retail selling price for each of the following items.

ITEM	WHOLESALE VALUE	MARKUP RATIO	RETAIL SELLING PRICE
a. 2-bloom carnation corsage	$ .85	____	____
b. mound arrangement	5.60	____	____
c. decorated plant	8.50	____	____
d. 8 carnations (cash-and-carry)	2.40	____	____
e. contemporary freestyle arrangement	5.80	____	____

2. What would be the selling price for an arrangement with a wholesale value of $8.25, using divisional pricing?

3. A customer orders a $35.00 sympathy spray. Determine the wholesale value of the flowers and materials that can be used in the spray using the ratio markup method.

4. Calculate the retail selling price of a basic arrangement using the materials listed below by three different pricing strategies.

ITEM	WHOLESALE VALUE
container	$ .90
flowers and foliage	7.25
miscellaneous supplies (cardett, enclosure card, etc.)	.25

PRICING STRATEGY	RETAIL SELLING PRICE
Standard Markup Ratio	_____
Retail Cost of Goods Plus Labor (25 percent)	_____
Divisional Percentage Pricing	_____

B. Short Answer Questions

1. What is meant by leader pricing?
2. What is meant by nesting pricing?

Selling in the Flower Shop

Terms to Know

clearinghouse
greeting approach
merchandise approach
selection guide
service approach
suggestion selling
upselling

Materials List

*blank order forms
  from a flower shop*
practice telephone

OBJECTIVE

To develop skills in retail sales.

Competencies to Be Developed

After completing this unit, you should be able to:

- identify characteristics of a professional salesperson.
- list the steps a professional salesperson goes through in helping a customer make a purchase.
- demonstrate effective selling skills.
- demonstrate telephone sales techniques.
- identify procedures for handling wire orders.

Introduction

Why do people drive past one flower shop to trade with another? Is it because the other store has a better quality product or more creative designers? The answer to these questions is usually no. Most good flower shops have a quality product and good designers. The answer lies in the treatment people receive.

The customer is the lifeblood of the retail flower shop. The reception, treatment, and satisfaction of the customer will determine the success of the business. Having the best designers will not ensure the success of a shop if customers are treated poorly. As a florist, you are not only selling flowers, but service and convenience.

CHARACTERISTICS OF A PROFESSIONAL SALESPERSON

A Friendly, Helpful Attitude

A friendly, helpful attitude is communicated to customers by a number of personality traits. Among these are sincerity, enthusiasm, empathy, poise, and perceptiveness.

Sincerity means having a genuine interest in the customer. Get the customers talking so that you can learn more about their needs. Listen attentively to what they say so they see you are interested in them.

All successful sales people have enthusiasm about the product they sell. Express that feeling to the customer. People are more likely to buy an arrangement of flowers if they feel the salesperson really likes it.

Having empathy means you are able to see things from the customer's point of view. When you do this, you will be better able to identify the customer's buying motives and help make a suitable selection.

Poised sales people are generally self-confident and have self-control. They remain calm under pressure, and they are tactful.

Being perceptive means observing your customers for traits that might not be obvious at first. The tone of voice people use and their actions while in the shop are clues that the salesperson can use in dealing with the customer. The sales tactics that you use with a shy individual would be different from the tactics you would use with an open, relaxed customer.

Thorough Product Knowledge

Successful sales people learn all they can about the product they sell. Many customers have little knowledge of flowers and their cost. They depend, to a large extent, upon the salesperson to provide them with the appropriate flowers for the purpose intended. If you are to help the customer make such decisions, then you have to appear confident in your knowledge about flowers (Figure 23-1).

FIGURE 23-1
Salespersons should know the product being sold so they can help customers make wise decisions. *Photo courtesy of M. Dzaman*

On a daily basis, be familiar with what is available for sale. Get to know the names of flowers, where they were grown, and any special information that would help the customer care for them. Also know the prices of various flowers so you can make suggestions in various price ranges. If a customer asks for a specific flower that you do not have, be able to offer a good substitute. A customer will feel good about a purchase if you have shown confidence in helping him make a decision.

Effective Selling Skills

Effective selling skills have to be learned. Professional sales people go through a number of steps in helping a customer make a purchase.

Step 1. *The Initial Approach or Greeting* Generally three methods may be used to approach customers in the retail florist: the service, greeting, and merchandise approaches.

The **service approach** is best used when it is obvious the customer is in a hurry. In this approach, the salesperson might ask the customer if he or she needs assistance by asking, May I help you? In many situations this approach is not effective because it usually results in a negative response such as, I'm just looking. When this happens, you lose control of the sales situation. If customers are in doubt about a purchase, they may leave without giving you a chance to help them with the decision.

In the **greeting approach**, the salesperson simply welcomes the customer to the store. Use the customer's name if possible with an appropriate greeting, such as, Good morning. If you know the person, make a personal remark if possible. This approach begins conversation and you can begin to discover the customer's needs.

In the **merchandise approach** the salesperson makes a comment about an item that has the customer's attention. To use this approach, The customer must stop to look at a specific item (Figure 23-2). This approach is usually the most effective because it focuses attention on the merchandise and gives you the opportunity to tell the customer something about it.

Step 2. *Determining Needs*   Customer needs are directly related to buying motives. The American Floral Service, Inc. has identified three reasons why customers buy flowers:

1. To fulfill an emergency need.
2. For self-satisfaction and personal enjoyment.
3. To receive personal recognition.

FIGURE 23-2

One approach to getting a customer is to make a comment about an item that has the customer's attention.

The main reason people buy flowers is for personal recognition. The salesperson's job, therefore, is to discover the level of recognition the customer is seeking. By listening and asking probing questions, this information can be obtained. For example, a man sending flowers to an important client is seeking a high level of recognition. In order to help him make a purchase, you need this information.

Step 3. *Product Presentation*   After you have determined the reason a customer is buying flowers and the level of recognition the customer is seeking, you should be able to select a few items that might match those needs. Base your suggestions on product, emphasizing the benefits. Include, but do not emphasize, price.

To avoid overwhelming the customer, show no more than three products at a time. If you are not sure about price, show one item from three different price ranges. For example, you might show arrangements for $20.00, $25.00, and $30.00. Explain the added benefits of the higher priced item. People are usually willing to pay more if they are assured of the value of the item. This procedure is called **upselling** and often results in a larger sale. Remember to emphasize product, not price. If possible, get the customer involved in the sale. Encourage the customer to hold or smell specific flowers, rather than looking at them in a cooler (Figure 23-3).

Use descriptive language to explain how an arrangement will meet the customer's needs. Avoid such words as "nice" and "cute," using terms like "fragrant" and "lasting quality," instead.

To present a product well, you must be well informed. This requires effort. Make notes each morning about the flowers on hand and product ideas to use in sales presentations.

Step 4. *Handling Questions and Objections*   Objections and questions should be viewed as positive feedback. Answering questions gives you the opportunity to present more information. If a customer seems to like one arrangement but has some reservations about making a decision, suggest some changes that could be made to better suit the customer's needs. If the price range of the arrangements you have shown is too high or too low, put those aside and show other arrangements in the correct price range. Show empathy toward the customer and offer product and price alternatives.

FIGURE 23-3

Get the customer involved in the sale by holding or smelling flowers. *Photo courtesy of M. Dzamen.*

Step 5. *Closing the Sale* This calls for the customer to make a decision. When you ask for a buying decision, base the decision on product, not price. Ask, for example, if the customer would prefer the arrangement of lilies or the arrangement of mixed spring flowers.

If the customer is still not ready to make a decision and offers objections to both arrangements, quickly regroup and make suggestions that are more appropriate to his or her needs.

Step 6. *Suggestion Selling* This encourages the customer to buy additional merchandise related to the original purchase. For example, a customer has purchased an arrangement for a birthday party. You might suggest balloons to go along with the arrangement. This is a good way to increase sales.

Step 7. *Writing up the Order* This order should include a brief description of the arrangement. It is a good idea to repeat the description to the customer so that there are no misunderstandings. Confirm the delivery date and the message to be included on the card. Itemize any additional charges such as delivery and sales tax. Give the customer the total charges.

If the customer does not have an established charge account, ask the customer how he would prefer to pay for the order.

Step 8. *Reassuring and Thanking the Customer* This is the last step of a sale. Help the customer feel confident that he or she has made a wise purchase. Thank the customer for selecting your shop and ask the customer to call again.

SELLING BY TELEPHONE

In many retail flower shops, a good telephone sales presentation may be more important than direct customer selling. Many of the orders received in the flower shop are taken by telephone. The same principles of selling already discussed in this unit apply to the telephone as well. The main difference is that the telephone salesperson must rely on words to describe the flowers and make the presentation. For this reason, the telephone salesperson must be knowledgeable about flowers, prices, and services offered by the shop. In the shop, a salesperson can show an arrangement. Over the telephone, the salesperson must describe it.

The person selling flowers by telephone must have a telephone voice which is easily understood and sounds friendly. Remember, the telephone customer can not see your smile. He or she has to hear it.

The Telephone Sales Presentation

Here are some important points to remember when selling over the telephone. These suggestions were summarized from "Selling With a Vital Difference," a video training series presented by American Floral Services, Inc.

1. *Answer Promptly.* Answer by the second ring (Figure 23-4). A short wait on the telephone seems much longer than a short wait in the flower shop.
2. *Greet the Customer.* Greet the customer and give the name of your flower shop and your name.

FIGURE 23-4

Answer the telephone by the second ring. *Photo courtesy of M. Dzamen*

"Good Morning. Sunny Side Florist. This is Nan Clark speaking."

3. *Use the Customer's Name.* The person calling will usually respond with the purpose of the call. If the customer does not give a name, make a simple service statement and ask the customer's name.

"I would be happy to help you select an arrangement. With whom am I speaking?"

4. *Ask for Delivery Information.* Ask for the appropriate delivery information (name and address of the recipient and delivery date).

5. *Ask Probing Questions.* Ask for the message on the enclosure card, or ask a probing question about the recipient. This will help you determine the reason the customer is buying the flowers.

6. *Offer Product and Price Presentation.* If the customer is not specific when placing the order, present three specific product suggestions. Remember, you must visualize and suggest specific arrangement ideas based upon the flowers you have in the shop. Price should be a natural part of the product presentation, but should not be emphasized.

7. *Ask the Customer for a Buying Decision.* Ask the customer for a decision based on product selection, not price. Offer alternatives if the customer has objections.
8. *Confirm the Order.* Repeat a brief description of the arrangement. Confirm the delivery date and details, as well as the message to be included on the card.
9. *Summarize the Charges.* Itemize any additional charges such as those for delivery and sales tax. Give the customer the total price.
10. *Determine the Method of Payment.* Ask how the client would like to pay for the order.
11. *Reassure the Customer.* Give your customer reassurance that he or she made an appropriate selection. Confirm that the recipient will be pleased.
12. *Thank the Customer.* Thank the customer for using your shop and for his patronage. Ask the customer to please call again.

Remember, professional selling requires preparation. It is not realistic to think you can answer the phone and hope the right ideas will come to mind. Selling requires a desire to be a professional and the discipline to make daily preparations to sell professionally. Know what is available for sale and be ready to suggest specific ideas and price ranges.

SENDING FLOWERS BY WIRE

Most florists across the country belong to at least one wire service (Figure 23-5). Many belong to several. A wire service is a **clearinghouse** for floral orders. The clearinghouse is essentially a bookkeeping service which ensures the sending and receiving florist that each shop will receive proper payment for the orders transmitted and delivered. On a basic level, a wire service extends credit to member florists and serves as a communication link between them.

An example will help you to understand how a wire service operates. Let's say you want to send flowers to someone in St. Louis, Missouri. You call or visit your local florist and place an order for a $30.00 arrangement. Your local florist takes all the information, including the name and address of the recipient. You pay your local florist.

The local florist uses a wire service membership directory to find a florist in St. Louis who will take the order, make the

WIRE SERVICES	
American Floral Services P.O. Box 12309 Oklahoma City, OK 73157-2309 800-456-7890	Carik Services, Inc P.O. Box 24286 Denver, CO 80224 800-692-6936
Florafax, International 4175 South Memorial Drive Tulsa, OK 74145 918-622-8415	Florists' Transworld Delivery 29200 Northwestern Hwy. P.O. Box 2227 Southfield, MI 48037 313-355-9300
Redbook Florist Services P.O. Box 258 Paragould, AR 72451 800-643-0100	Teleflora 12233 W. Olympic Los Angeles, CA 90064 213-826-5253 800-321-2654

FIGURE 23-5

Major wire services.

arrangement, and deliver it. But you paid your local florist and the St. Louis florist is the one doing all the work. This is where a wire service enters the picture.

The local florist, called the *sending florist,* receives a commission for taking the order. This amount is usually about 20 percent. The florist in St. Louis is called the *receiving florist* and receives 73 percent of the value of the order. The receiving florist reports the order to the wire service who pays them while sending a bill to the sending florist for 80 percent of the total amount. The wire service keeps 7 percent of the order for advertising and clearinghouse charges.

For a $30.00 order, the clearinghouse receives $2.10, the sending florist receives $6.00, and the receiving florist is paid $21.90. Even though the receiving florist only receives 73 percent of the order, they are expected to deliver 100 percent of the value of the order.

Benefits of a Wire Service

1. You can accept an order for delivery to any location, worldwide.
2. You receive orders from locations other than your local trade area.

3. You are given the opportunity to deliver to a potential new customer.
4. Educational services are offered to member florists by several of the wire services.
5. Design shows and state and local floral functions are assisted by all of the wire services through the sponsorship of designers and participation in trade shows.
6. National advertising supplies information to the public and helps persuade consumers to purchase flowers.
7. You can subscribe to magazines published by the wire services. Some of the more familiar magazines are *Florist* (FTD), *Flowers* & (Teleflora), *Design for Profit* (Florafax), and *Professional Floral Designer* (AFS).

Using a Wire Service

When taking a wire order, the florist needs the same information that is collected for any other order. However, there are differences in the way the information should be taken and the customer's options described.

First, the florist may suggest that the customer look at a wire service **selection guide** which has color pictures of a wide variety of arrangements. The selection guide is used as a means of communicating design styles and customer preferences to the florist receiving the order (Figure 23-6). Each

FIGURE 23-6

A retail florist uses a wire service selection guide to assist a customer in wiring flowers.

design has a code number. When the order is given to the receiving florist, the code number is all that is necessary to describe the design wanted. The florist simply locates the code number in his own selection guide and designs a matching arrangement. When taking a wire order from a customer, it is often necessary to ask for a second and possibly a third choice, in case the receiving florist cannot provide the first choice.

Pricing is slightly more complicated when dealing with a wire order. An arrangement selling for $30.00 in one town may sell for $40.00 in another town. This is often true of orders being transferred from small towns to large cities. An experienced florist will learn the price levels of different areas and be able to deal with the pricing difference by informing customers or by marking up the prices of arrangements.

The florist then selects a receiving florist from the wire service membership directory. The code number of the shop is recorded and the order is placed by telephone. The customer's bill includes the cost of the arrangement plus telephone or transmittal charges.

In this unit, we have discussed the benefits of being a member of one or more wire service agencies. Being able to please a customer is probably the greatest benefit. If a florist is not a member of a wire service, then the customer will go elsewhere. The florist has lost a sale and perhaps a regular customer.

SELLING OVER THE INTERNET

Many flower shop owners have recognized the potential of the Internet to expand their customer area. Over the Internet, a shop's customer area can become worldwide. By establishing a home page on the Internet, the local florist can advertise items available for sale to customers anywhere the Internet exists.

Flower shops are currently using the Internet in a variety of ways. Some shops scan pictures of wire service selections onto their home page. Customers make their selection just as they would in the flower shop. Some shops provide a toll-free number that customers call to place their orders. Others have an order form which customers complete and the order is sent to the flower shop by e-mail. The order form asks for the same information that would be completed for an order placed in the shop.

The floral order is handled just as any other order. If the order is in the delivery area of the flower shop, it is completed and delivered in the daily work of the shop. If the order is out of the shop's delivery area, the flower shop becomes the sending florist and places the order with a florist in the area of delivery. Thus the order is handled just as any other wire service order.

Some florists advertise their own designs on their home pages and promise delivery anywhere in the continental United States. For example, one flower shop in Hawaii advertises tropical arrangements and guarantees their arrangements. The customer simply makes a selection from the home page, completes the billing and delivery information, and e-mails the order to the shop in Hawaii. The shop completes the order and has it delivered by one of the express delivery services. The customer has to pay extra for the delivery service, so this must be added to the cost of the flowers.

Many shops have successfully expanded their productivity by using the Internet. Selling over the Internet is known as e-commerce and is rapidly expanding. If you would like to know more about selling flowers over the Internet, type flowers.com onto your favorite search instrument and browse the many web sites that will be provided.

Student Activities

1. As a class activity, make up situations in which an individual needs flowers for a specific purpose. Select some class members to play the customer and others, the retail florist salesperson. Evaluate the presentation of the student acting as salesperson.

2. Use role-playing situations such as those described in activity 1 for a telephone order. Make copies of a local florist's order form and use toy telephones. Position the two students back-to-back while the role playing is taking place.

3. Invite a local florist to talk to the class about wire services.

4. Ask your instructor to make copies of the sample order form (Figure 23-7). Complete the order form for a $35.00 freestyle arrangement using the following information.

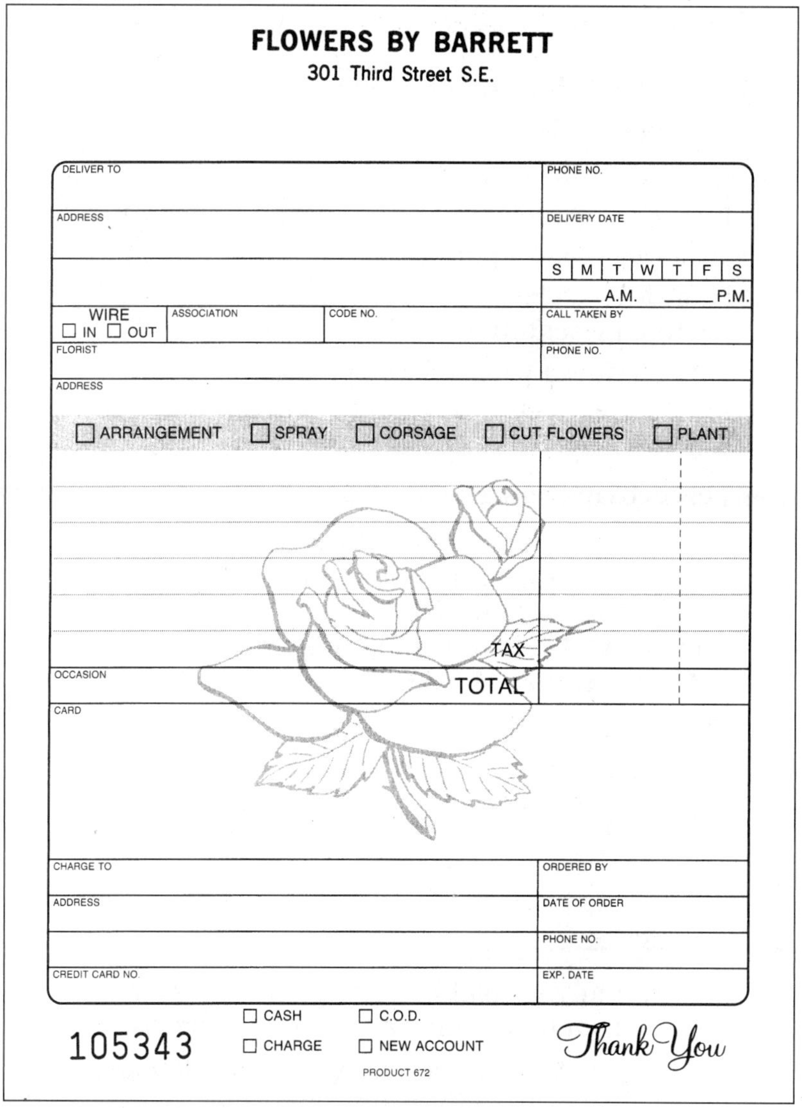

FLOWERS BY BARRETT
301 Third Street S.E.

DELIVER TO	PHONE NO.
ADDRESS	DELIVERY DATE

	S	M	T	W	T	F	S

_____ A.M. _____ P.M.

WIRE ☐ IN ☐ OUT	ASSOCIATION	CODE NO.	CALL TAKEN BY
FLORIST			PHONE NO.
ADDRESS			

☐ ARRANGEMENT   ☐ SPRAY   ☐ CORSAGE   ☐ CUT FLOWERS   ☐ PLANT

TAX

OCCASION   TOTAL

CARD

CHARGE TO	ORDERED BY
ADDRESS	DATE OF ORDER
	PHONE NO.
CREDIT CARD NO.	EXP. DATE

105343   ☐ CASH   ☐ C.O.D.
☐ CHARGE   ☐ NEW ACCOUNT   *Thank You*
PRODUCT 672

FIGURE 23-7
Sample order form.

Person placing order:	Mr. Jack Hill 2700 Misty Circle Moultrie, Georgia 31768
Deliver order to:	Mrs. Jack Hill same address
Information to be placed on card:	Happy Anniversary Love, Jack
Date of delivery:	February 27

Self-Evaluation

A. Provide a brief answer for each of the following.

1. List the characteristics of a professional salesperson.

2. List the steps a salesperson goes through in helping a customer make a purchase.

3. How does a telephone sales presentation differ from an in-store presentation?

4. Explain how a wire service operates.

5. What benefits do wire services offer florists?

6. How can the Internet be used to increase a shop's productivity?

B. Calculate the amount of money that each of the following would receive from a $50.00 order placed by a customer to be wired to another town.

1. Sending florist _____

2. Receiving florist _____

3. Wire service _____

C. True or False

_____ 1. Only a small percentage of the orders received by a retail florist are over the telephone.

_____ 2. When a salesperson is taking a telephone order, being friendly is not important as long as the salesperson is efficient and obtains all needed information.

_____ 3. A salesperson must be knowledgeable about flowers as well as good sales techniques.

_____ 4. Because customers usually know what they want when calling a florist, little training is needed to be a salesperson in a flower shop.

_____ 5. The service approach is best when greeting a customer who is in a hurry.

———— 6. The greeting approach is used when a customer is examining a specific item for sale.

———— 7. People buy flowers to gain recognition.

———— 8. Objections and questions from a customer should be viewed as negative feedback.

———— 9. Always repeat a description of an order to the customer.

———— 10. A wire service is a clearinghouse for floral orders.

Displays

Terms to Know

artistic displays
buying up
display
display elements
impulse buying
product-oriented
displays
theme displays
visual merchandising

OBJECTIVE

To plan a visual display.

Competencies to Be Developed

After completing this unit, you should be able to:

- distinguish between visual merchandising and display.
- list the four primary purposes of a display.
- list the secondary purposes of a display.
- state the purpose of artistic displays.
- list the two categories of display and give examples of each.
- list five display arrangements.

Introduction

The goal of any florist is to sell products or services. In order to do this, customers must be attracted to the store. The manner in which a store accomplishes this is known as **visual merchandising.** Visual merchandising includes building design, store layout, lights, signs, fixtures, displays, and general store decorations so that the right image is projected to the customer (Figure 24-1). In this unit, we will deal with only one component of visual merchandising, that of **displays.** A

FIGURE 24-1

Visual merchandising begins with the store exterior. *Photo courtesy of M. Dzamen*

display is the visual and artistic aspect of presenting a product to the customers.

Customers have a wide choice of outlets from which to purchase flowers. The goal of the retail flower shop is to portray an image of uniqueness which attracts the customer. An advertisement might be effective in attracting customers to a shop, but once they arrive at the shop, the image created by the shop exterior and the displays must create the mood or feeling implied by the advertisement.

A customer's opinion of a shop begins with the store exterior. A shop should be attractive and well maintained with easy-to-read signs. The shop should have adequate parking with easy access to the store entrance. Customers are not attracted to a shop that is poorly maintained and unattractive. This creates a negative image and often leads to a loss of customers.

PRIMARY PURPOSES OF DISPLAYS

The primary objective of any display is to motivate customers to make purchases. The ability to do that requires an understanding of the way visual merchandising works.

An effective visual display will achieve the four primary purposes of display and will often serve other purposes as well. The four primary purposes of displays are to (1) attract attention, (2) arouse interest, (3) create desire, and (4) initiate buying.

Attract Attention

The consumer must be attracted to the store or merchandise before a sale can be generated. A display attracts attention by causing people to look. If this function is not adequately performed, none of the other purposes are likely to occur.

Attracting attention to a display can be achieved in a number of ways. Color, contrast, light, and motion can be used to attract attention. Motion devices include turntables, fountains, fans, and flashers. They are readily available and inexpensive.

Something as simple as the use of fabric brings visual attention to a display but there should be a specific reason for incorporating fabric into the display. The fabric can make a dramatic color statement that pulls visual attention to the display or help to separate items or colors within the display.

Arouse Interest

Attracting attention to a display is not enough. The display must hold the viewer's attention long enough to develop interest in the merchandise. Something in the display must act as a focal point. Starting from this point, the viewer should move visually through the display, taking in all the selling points the display is emphasizing.

Create Desire

In the process of looking at the display, the viewer should become enthusiastic about buying the merchandise. When products are shown in everyday settings, the customers can envision how they might use the product in their own home (Figure 24-2). When products are displayed in a glamorous or fantasy setting, the customer can become a part of the fantasy by purchasing the products shown.

FIGURE 24-2

A display can show how merchandise can be used in the home.

Initiate Buying

If the first three steps of selling have been accomplished, buying action should result. The display should aid the customer in taking this last important step. Information such as the price of merchandise is usually needed at this point (Figure 24-3). Additional information, such as benefits or uses of an item, may also be helpful.

SECONDARY PURPOSES

While the primary purposes of a display should always be achieved, a number of secondary purposes can be also served by window and in-store displays. These include:

1. Encouraging **impulse buying.** A customer sees an item and buys it even though the item was not what the customer originally intended to buy.

FIGURE 24-3

A display should furnish the customer information about merchandise, such as the cost of the item. *Photo courtesy of M. Dzaman*

2. Encouraging "**buying up.**" A customer buys a more expensive item than planned.
3. Encouraging multiple sales. A customer purchases accessories or unrelated items.
4. Creating store image. Creating or changing a shop's image can be done with visual displays.
5. Showcasing products. Displays help customers associate the products available with the flower shop itself.
6. Reinforcing advertising. Whenever possible, a visual display should complement a shop's advertising plan.
7. Educating consumers. Displays which provide educational messages are effective because they take advantage of the consumer's natural desire for knowledge.
8. Encouraging future sales. At a later date when a gift is needed, a customer may remember a displayed item and return to the shop.

ARTISTIC DISPLAY

Florists sometimes use **artistic displays** in the flower shop (Figure 24-4). These displays are designed to stop traffic and focus attention. An artistic display does not sell directly un-

FIGURE 24-4
Artistic displays stop customer traffic and capture attention.

FIGURE 24-5
A seasonal display is one type of theme display.

less the display is coordinated with an adjacent stock of the same merchandise. Artistic displays can be used to pull customers through a store, to draw people to a "dead" area, to stop shoppers at an important section, and to create an image.

CATEGORIES OF DISPLAY

Most displays can be categorized as **theme displays** or product displays. Each can be equally important in advancing the store's sales and image.

Theme Displays

Theme-oriented displays are based on a specific subject or topic. Some of the more common themes for floral displays are:

1. Holidays—Christmas, Valentine's Day, Easter, etc.
2. Seasons—spring, summer, autumn, and winter (Figure 24-5).
3. Life events—weddings, graduations, proms, etc.
4. Commemorations—Veterans Day, Secretaries' Day, etc.
5. Community celebrations—local fairs, civic projects, political and cultural events.

A suggested monthly planner for display themes can be found in Figure 24-6.

January
Winter Seasonal Display
Valentine's Day

February
Valentine's Day
Presidents' Day
Mardi Gras
National FFA Week

March
Spring Seasonal Display
St. Patrick's Day
Easter

April
Spring/Easter
Passover
Secretaries' Week

May
Mother's Day
Summer Seasonal Display
Memorial Day
Commencement

June
Summer Seasonal Display
Father's Day

July
Summer Seasonal Display
Independence Day

August
Summer Seasonal Display
Back to School

September
Labor Day
Back to School
Grandparents' Day

October
Autumn Seasonal Display
Halloween
Bosses' Day
Mother-in-law's Day
Sweetest Day

November
Autumn Seasonal Display
Veterans' Day
Thanksgiving

December
Winter Seasonal Display
Christmas

FIGURE 24-6

Flower shop display—monthly planning calendar.

Product-oriented Displays

Product-oriented displays focus on the direct promotion of merchandise. These are grouped according to the following:

1. Single-item display. Emphasizes one piece of merchandise (Figure 24-7).
2. Line-of-goods display. Emphasis on a single type of merchandise.
3. Related merchandise. Emphasis on a collection of items that are similar or used together (Figure 24-8).
4. Variety displays. Emphasis on unrelated merchandise that may only have color or price in common (Figure 24-9).

FIGURE 24-7
A single-item display.

FIGURE 24-8
Related merchandise display.

FIGURE 24-9

A variety display contains a group of unrelated arrangements.

Product-oriented displays may place emphasis on new or old product lines, giftware, plants or flowers, or accessory lines.

DESIGNING DISPLAY ARRANGEMENTS

Display elements are the components used in assembling displays. These elements can be mixed in many combinations to create the desired effect. The merchandise to be displayed, fixtures, and props are all display elements.

When designing a display, keep in mind the principles of design discussed in Unit 3. These same principles can be used to guide you in planning displays, but will have to be applied in a slightly different manner.

Among the most commonly used display arrangements are repetition, step, zigzag, pyramid, and radiation arrangements.

FIGURE 24-10
Repetition arrangement.

Repetition Arrangement

This is simple in form and achieves its effect through the repetition of similar items (Figure 24-10). Height, spacing, and direction are the same for all items.

Step Arrangement

Merchandise of varied sizes is often displayed in a progression from smallest to largest. This creates a stair-step pattern which is easy for the eye to follow (Figure 24-11). The difference in height from one level to the next should be the same, and all items should face in the same direction.

FIGURE 24-11
Step arrangement.

FIGURE 24-12
Zigzag arrangement.

Zigzag Arrangement

A zigzag arrangement of merchandise leads the eye back and forth and up and down a display. It provides an interesting and continuous line for the eye to follow (Figure 24-12). Columns or pedestals can be used to achieve the height desired to create the zigzag pattern.

Pyramid Arrangement

The pyramid is a triangular arrangement of products with a broad base rising to a center peak (Figure 24-13). This is a formal arrangement and is effective with most merchandise. Potted flowers and plants are often displayed on a pyramid-shaped island.

FIGURE 24-13
Pyramid arrangement.

Radiation Arrangement

In the radiation arrangement, merchandise appears to radiate from a central point (Figure 24-14). Columns and pedestals can be arranged in a display at appropriate heights to create a fanlike arrangement of merchandise or a complete circle of merchandise radiating from a central point.

The display arrangements discussed in this unit are in their most simplified forms. Most displays consist of one or several of these arrangements in a single display.

FIGURE 24-14

Radiation arrangement.

Student Activities

1. Draw a layout of an in-store display using one or more of the display arrangements.

2. If props are available, construct the display you designed in some area of your classroom or school.

3. If your school has a marketing education program with a display window, ask the instructor if your floral design class might use the display window. Plan and construct a window display as a class project.

4. Use a cardboard box as an enclosed window and construct a small window display.

Self-Evaluation

A. Provide a short answer for each of the following.

1. Distinguish the difference between visual merchandising and display.
2. List the four purposes of display.
3. Briefly explain how a display can attract attention.
4. List the secondary purposes of display.
5. What is the purpose of an artistic display?

B. Identify which of the following displays are theme-oriented and which are product-oriented.

1. Valentine's day
2. wedding
3. a new line of vases
4. a cultural event
5. a display of cut flowers
6. Secretaries' day
7. a display of potted plants, pots, and growing medium
8. a display of balloons

C. Illustrate each of the following display arrangements.

1. zigzag
2. step
3. radiation
4. repetition
5. pyramid

Delivery

Terms to Know

control board
delivery pool
holiday deliveries
last-minute deliveries
out-of-area deliveries
regular deliveries
zoning system

OBJECTIVE

To identify the components and the importance of a floral delivery system.

Competencies to Be Developed

After completing this unit, you should be able to:

- identify the job requirements for a delivery person.
- discuss the training needed to be a delivery person.
- identify delivery categories.
- explain the delivery process.
- tell how to load the delivery vehicle.
- plot a delivery route.
- role play a delivery.

Introduction

Delivery is a service that sets the retail florist apart from other retail services and especially from mass-market stores that sell flowers. Delivery is a personalized service and is an important selling point for the retail florist. Customers can usually buy flowers cheaper at a mass-market florist, but they cannot get the personalized delivery service offered by their retail florist shop.

The delivery personnel and procedures are so important that they can determine a shop's image and profitability. Delivery should involve training and organization.

THE DELIVERY PERSON

A person receiving flowers never has contact with the flower shop so the delivery person truly represents the shop. The appearance and actions of the delivery person contribute to the consumer's opinion of the shop. While this image is important in gaining new customers and retaining repeat customers, it is often overlooked by flower shops in their hiring of delivery persons.

Most shops have regular full-time or part-time delivery personnel. In small shops, other shop personnel may also serve as the delivery person. During peak seasons, extra help may be hired to assist with deliveries. Regardless of the type of delivery personnel a shop has, all need to go through a brief training period.

The appearance of the delivery person is a critical factor to consider (Figure 25-1). The beauty of flowers should not be lessened by a dirty or unkempt delivery person. Any person who comes in contact with customers should be well groomed and appropriately dressed for the job. Many shops

FIGURE 25-1

Delivery personnel should be well groomed and dressed appropriately for the job. *Photo courtesy of M. Dzaman* prefer uniforms. A uniform makes it easy to identify the person as a delivery person.

The delivery person needs to be well mannered and genuinely polite. Customers appreciate friendliness and concern for the flowers they are receiving.

Knowledge of the area is another factor to consider when hiring delivery persons. A person who has experience with local driving may be able to make deliveries more efficiently. Also, check the person's driving record. Find out if there have been any tickets or accidents.

A person's ability to speak, read, and write properly is critical. In some areas, bilingual ability may also be important. Filling out a job application form, delivery form, and reading a map are ways of checking for these abilities.

Training

If delivery personnel are to do a good job, they need to be aware of their responsibilities. A thorough job description which itemizes duties is important. The duties, which are explained during the formal training period, can include in-shop responsibilities as well as delivery duties. The training period should include the following:

1. A review of rules and regulations which apply to all shop employees.
2. Protocol for the care and maintenance of the delivery vehicle.
3. An explanation of the ordering and delivery process.
4. Product information, such as basic care and handling practices, flower and plant identification, and terminology.
5. How to package flowers and load the delivery vehicle.
6. How to schedule and route deliveries (Figure 25-2).
7. Review of rules and regulations for making deliveries to hospitals, funeral homes, and churches.
8. A trial delivery in which the shop manager observes the delivery process and the individual's communication skills.
9. How to handle paperwork.
10. Demonstration of in-shop responsibilities to perform when not making deliveries.

FIGURE 25-2

Plotting a delivery route is part of the training process for a delivery person. *Photo courtesy of M. Dzaman*

THE DELIVERY VEHICLE

The selection of a delivery vehicle is an important decision for the florist (Figure 25-3). The delivery vehicle represents a major expense that must be carefully controlled. The vehicle is also a reflection of the shop's image. Because the vehicle is constantly seen throughout the community, it should be kept clean and in good repair. An attractive easy-to-read logo giving the shop's name should be on the vehicle. This serves as a major source of advertisement for the shop.

FIGURE 25-3

The delivery vehicle is a reflection of the shop's image in the community. *Photo courtesy of M. Dzaman*

DELIVERY CATEGORIES AND CHARGES

Delivery is a service offered by most retail florists, but the expense of buying and maintaining a delivery vehicle and hiring delivery personnel makes it a costly item. This expense must be passed on to the customer. Most florists charge for delivery service. Exceptions are usually funeral home and hospital deliveries. Many florists do not charge for these because of the large volume of flowers going to these locations. The cost is usually absorbed by charging for other deliveries.

Charges for this service usually depend upon the type of delivery. The following is a list of categories that describes the needs of most florists.

1. **Regular deliveries.** These are deliveries handled through the daily deliveries within the shop's established delivery area.
2. **Out-of-area deliveries.** These locations are beyond the shop's normal delivery area. The orders may be transferred to another florist near the customer or may be serviced for established customers or special circumstances.
3. **Last-minute deliveries.** These are emergency deliveries requiring immediate, special service.
4. **Holiday deliveries.** These occur during holiday or peak times when extra personnel are hired and more special requests made.

Most florists charge for all deliveries and charge extra for special service deliveries. Many florists establish a base charge for regular deliveries depending on the cost per mile for the delivery vehicle. The base charge would cover deliveries within a specific range of miles. Beyond this range, the delivery charge would increase. The florist can draw circular grids on a map of the delivery area and establish charges accordingly. This process also allows sales people to look at the map and determine delivery charges at the time the order is placed. This eliminates any misunderstandings with the customer.

Other special delivery requests can be based on regular delivery charges plus an amount determined by the florist. Timed deliveries might be double or more the normal charges.

PROCESSING AN ORDER FOR DELIVERY

Many times delivery problems can be traced to the salesperson taking the order. Complete and accurate delivery information is just as important as the design and billing. Remember, you cannot charge a customer for an order that was not delivered.

One method of processing an order is to use multiple-copy order forms. This allows a copy to be sent on delivery while the control copies remain in the shop.

A numbering system on all invoices is also important. It allows for quick and easy reference in locating a problem order. When using this system, always place the invoice number on the back of the enclosure card sent with the arrangement. If there is a problem with the order, the customer can quickly give the florist the invoice number.

Large florists may use a control board and zoning system to ensure quick and efficient deliveries. A **control board** is a visual control over the day's deliveries. Each delivery ticket is put on the proper day's hook so that the delivery person knows in advance what orders must go where on what day. A control board might be used to organize deliveries by zones, with a peg or hook for each zone.

A **zoning system** groups orders together according to delivery runs. Areas of the shop are marked off according to delivery zones with all ready-to-go orders placed in the proper zone. A special delivery should be marked with some type of distinctive marking so that it can receive special attention.

LOADING THE DELIVERY VEHICLE

Before loading the delivery vehicle, the driver or another designated person should decide upon the quickest, most efficient route. This can be determined from the list of invoices. Many drivers mark the route on a city map. A laminated map and water solvent markers make this an easy process.

Speed and efficiency of the delivery run are partially determined by the manner in which the delivery truck is loaded. The driver should arrange the items in the order in which they will be delivered. The last item to be delivered should be loaded first and placed in the rear of the delivery vehicle.

FIGURE 25-4

When loading the delivery vehicle, use items to buffer, anchor, and support the flowers. *Photo courtesy of M. Dzaman*

Use items to buffer, anchor, and support the arrangements (Figure 25-4). These can include boxes filled with crushed newspaper or packaging material, sandbags, and support racks. Commercial products, such as cardboard bases, foam trays, and anchor boards, can also be purchased to help support and protect flowers during delivery.

MAKING THE DELIVERY

The delivery person may have several types of deliveries to make on a route. These may include hospitals and funeral homes. The driver should be familiar with the policies regarding deliveries at each of these. Funeral homes may have cut-off times for deliveries. These should be noted on the invoice and tagged accordingly.

Many hospitals have a central delivery room for flowers where they wait until volunteers deliver them to the rooms. The delivery person needs to know the policies at each hospital where deliveries are made.

When making deliveries to offices and family dwellings, use the front door. Be prepared with a delivery speech such as a greeting, quick tips, and a closing. Be prepared to answer questions about the flowers. Remember, flowers are special so deliver them with that in mind. Friendly, helpful hints may lead to repeat business or a new customer.

FIGURE 25-5
Keep a repair kit in the delivery vehicle so simple repairs can be made on the delivery route. *Photo courtesy of M. Dzaman*

When the recipient is not home, and the flowers are left elsewhere, a tag should be left advising where the flowers have been taken. A tag should be left on each door to ensure that the notice is seen. If flowers cannot be left elsewhere, the tag will advise the recipient to call the shop.

Keep a delivery kit in the delivery vehicle (Figure 25-5). This kit should include local phone books, street guides, a city map, pencils and pad, and basic flower repair tools. The repair kit should include a cutter, wires, pins, and picks. Repairs made on the route will reduce the number of arrangements returned to the shop and prevent a second delivery from being made.

Upon returning to the shop, the delivery person should fill out a report form on any deliveries returned indicating the reason for the nondelivery. This will aid in follow-up.

DELIVERY POOLS

An alternative to in-house delivery is a multishop **delivery pool.** This involves a group of shops pooling delivery resources and personnel. Each shop brings its delivery items to a central meeting point by a specified time. The deliveries are divided by areas, and each shop then takes all of the deliveries for its area.

Advantages of a delivery pool include saving time and money due to less driving. The disadvantages are that the individual shop loses control of the delivery once an item leaves the shop. Also, extra packaging may be required because of the transfers involved.

Student Activities

1. Invite a flower shop owner-manager to talk to the class about the shop delivery system.
2. Laminate a city or county map and plan a delivery route from invoices provided by your teacher.
3. Role play making a delivery.
4. Explain how a delivery pool works.

Self-Evaluation

A. Short Answer Questions

1. Why should sales people be knowledgeable about the flower shop's delivery procedures?
2. Make a list of job requirements for a delivery person.
3. What training should a prospective delivery person receive?
4. Why is a good delivery service important to a florist?

B. True or False

_____ 1. Because of the expense, very few flower shops today offer a delivery service.

_____ 2. The delivery person often has in-store responsibilities as well as delivery duties.

_____ 3. The appearance of the individual hired as a delivery person is not important since he or she is not seen in the shop.

_____ 4. Communication skills are not important to the delivery person as long as he or she is a good driver.

_____ 5. Many florists do not charge for deliveries to hospitals and funeral homes.

_____ 6. When loading a delivery vehicle, the last item to be delivered should be placed in the delivery vehicle last.

_____ 7. A control board is used to organize deliveries.

_____ 8. A zoning system groups flowers to be delivered by other florists.

_____ 9. Cardboard boxes and sandbags may be used to support arrangements during delivery.

_____ 10. If the recipient of an order is not home, the flowers should always be returned to the shop.

Professional Organizations

OBJECTIVE

To identify professional trade organizations organized to assist the retail florist.

Competencies to Be Developed

After completing this unit, you should be able to:

- list several professional organizations open for membership to retail florists.
- identify membership requirements in the professional trade organizations.
- make a list of professional trade organizations.
- identify professional organizations with student membership categories.
- identify certification programs for retail florists.

Introduction

A student seeking a career in the retail florist industry should be aware that completing a training program at a high school, trade school, or college is just the beginning of a lifetime of learning (Figure 26-1). New techniques, styles, and business practices are continually developing. As a retail

430

FIGURE 26-1

A retail florist must continually learn to keep abreast of changes in the floral market.

florist, you will need to keep current on all the latest products and changes that are taking place in the industry.

Professional organizations, certification programs, and trade publications are ways that the retail florist can keep current on what is happening in the industry. These are supporting structures which provide leadership, guidance, and training, and they establish levels of excellence for the industry. The retail florist should be aware of these groups and gain membership in those that best serve his or her interest.

Trade publications help to educate and inform retail florists of the latest products and trends in the industry. Many publications contain articles on business practices and industry activities. Every retail florist should subscribe to one or more of these publications.

As a floral design student, you may want to join one of the professional trade organizations as a student member. Some of these also publish newsletters for students.

The purpose of this chapter is to help you in understanding the different types of trade organizations and publications and the purpose of each one. It serves as a directory of information and includes addresses and phone numbers.

TRADE ORGANIZATIONS

Trade organizations are usually nonprofit corporations. The purpose of the corporation is to assist its members through educational programs and to promote the industry. A trade organization may be established to meet the needs of a segment of the industry or the industry as a whole. For example, some organizations may be organized to benefit floral designers while others would benefit retail flower shop owners or growers. All organizations have information packets and membership applications available upon request. The following is a list of organizations and a brief description of each.

SAF (Society of American Florists)

The **Society of American Florists** is the only national trade association representing the needs and interests of all segments of the floral industry. Membership is open to growers, wholesalers, retailers, manufacturers, and suppliers of related products, educators, students, and other organizations in the industry. SAF keeps each informed of the needs and activities of the others.

The SAF provides a number of services to its members. Besides offering programs and workshops throughout the year, SAF provides several publications, newsletters, research reviews, videotapes, and printed materials to its members.

The SAF also has an effective government relations program that represents the interest of the members before congress and federal agencies.

In addition, the SAF sponsors a number of other organizations to provide support and recognition to industry members. Membership in SAF is easily available. Qualified businesses or individuals are only required to submit a membership application and application fee to SAF headquarters. The address and telephone number are listed in Figure 26-2.

TRADE ORGANIZATIONS IN FLORAL DESIGN AND MARKETING

American Institute of Floral Designers
 720 Light Street
 Baltimore, Maryland 21230
 (301) 752-3320

American Floral Marketing Council
 c/o Society of American Florists
 1601 Duke Street
 Alexandria, Virginia 22314
 (800) 336-4743

Society of American Florist
 1601 Duke Street
 Alexandria, Virginia 22314

PFCI—Professional Floral Commentators–International
 Same as the Society of American Florists

Redbook Master Consultants
 P.O. Box 1706
 3309 E. Kingshighway
 Paragould, Arkansas 72451

FIGURE 26-2

Trade organizations and their addresses.

AAF (American Academy of Floriculture)

The **American Academy of Floriculture** is a branch of the Society of American Florists. Its purpose is to encourage and recognize excellence in community and floral industry service. Membership in AAF is representative of all segments of the floral industry including growers, wholesalers, retailers, designers, educators, scientists, and allied businesses. Membership criteria include several high standards which must be met by each applicant. Because of these high standards, there is no student membership category in the AAF. Further information on membership requirements can be obtained by writing to the American Academy of Floriculture.

AIFD (American Institute of Floral Designers)

The **American Institute of Floral Designers** is dedicated to establishing higher standards in professional floral design. Membership is selective and based on rigid qualifications and demonstrated high professional ability. When AIFD appears after someone's name, it means that individual has achieved excellence in professional floral design.

AFMC (American Floral Marketing Council)

The purpose of the **American Floral Marketing Council** is to promote everyday sales of floral products. The AFMC is also sponsored by the Society of American Florists but is funded separately.

AFMC promotes what are known as nonoccasion sales. Their campaign spreads the word that "anytime is the right time" to buy flowers and plants. Retailers, as well as wholesalers and growers, participate in AFMC through their wire service. Dues are paid as 1 percent of all outgoing wire orders each month. The entire floral industry benefits from AFMC's promotional campaigns which encourage flower sales year-round.

PFCI (Professional Floral Commentators–International)

Professional Floral Commentators–International helps to educate all segments of the industry on innovations in floral design, floral marketing, merchandising, and fresh flower care and handling. The PFCI also introduces new products and floral varieties to the industry and seeks to improve commentating at all floral events. A commentator is an individual who explains and comments on floral designs at design shows. PFCI was established by the Society of American Florists. The letters PFCI behind a designer's name represent great achievement in floral commentating.

RMC (Redbook Master Consultants)

Redbook Master Consultants, Inc. is a nonprofit corporation organized to improve the floral industry through education. The main objective of RMC is to provide needed

FIGURE 26-3
Redbook Master Consultants provide needed educational programs to florists throughout the country.

educational programs to florists in the area where they live and work (Figure 26-3). Individuals working in the floral industry are eligible for membership upon meeting prescribed requirements of responsibility for a minimum of 3 years.

Allied Florists' Associations

Allied florists' associations are usually groups from within a region or a state. Each group acts as a support system for florists in its area. The advantage of allied florists' associations is that they focus on the needs and problems of the local florist, which may or may not be the same as a national organization. Membership is normally open to anyone. Interested individuals may contact a local wholesale florist for information on local allied associations.

State Florist Associations

Most states have a florist association devoted to the support and improvement of floral-related businesses within the state. The objectives of the **state florist associations** are similar to the allied florists' associations, but function on the state level. Most publish a newsletter to keep members informed of the association's activities. The state associations sponsor educational seminars and workshops on a variety of topics. Many also sponsor design competitions among the florists in the state. For information on membership in a state association, contact a local wholesaler.

Wholesalers' Open Houses and Design Schools

Wholesale florists are located throughout the country. These wholesale florists sell flowers and supplies to the retail florist in the area. Several times a year, many wholesale florists hold an open house and design school at which outstanding designers from across the country construct arrangements and share new products and ideas. Commentators present the designs to those present and share ideas with other designers (Figure 26-4). These shows offer retail florists the

FIGURE 26-4

A commentator presents floral designs at a wholesale florists' open house and design school.

chance to learn from each other and from selected outstanding designers. The dates for open houses can be obtained from the wholesale florists in your area.

TRADE PUBLICATIONS

Trade publications provide an important source of information to the florist. The information found in trade
magazines and newsletters is timely and up-to-date. These
publications provide business management tips, floral design techniques, information on the care and handling of
flowers, product information and sources, classified ads, and
news of upcoming events.

A partial list of publications available to florists is given
in Figure 26-5. A florist would probably want to subscribe to
a number of these and read them regularly to help ensure
that the most up-to-date ideas are being utilized in the
shop's operation.

STUDENT ORGANIZATIONS

A student who is serious about a career in retail floristry may
want to join one or several student organizations. These provide educational advancement and competition among its
members.

The National FFA Organization, an organization for students enrolled in agricultural education classes, has chapters
in schools across the nation where courses in floral design are
offered. The FFA has flower identification contests and design
competitions as well as leadership development (Figure 26-6).

FTD (Florists' Transworld Delivery Association) has a student membership category, offers design competition among
its student members, and publishes the FTD Future Florists'
Newsletter. The National Junior Horticulture Association has
an organization in each state and annually holds a national
convention. The NJHA provides opportunities for members to
display skills in identification, design, and marketing.

Many universities that offer degrees in horticulture
also sponsor a chapter of Pi Alpha Xi. This is a college-level
organization which provides flower-judging and design
competitions.

FLORAL INDUSTRY PUBLICATIONS

Design for People (quarterly)
Florafax International, Inc.
P.O. Box 470745
Tulsa, Oklahoma 74147
(918) 622-8415

Design with Flowers (monthly)
Herb Mitchell Associates, Inc.
234 East 17th St., Suite 202
Costa Mesa, California 92627
(800) 344-5995

Floral and Nursery Times (semi-monthly)
629 Green Bay Road
Wilmette, Illinois 60091
(708) 256-8777

Florist (monthly)
Florists' Transworld Delivery Association
29200 Northwestern Highway
P.O. Box 2227
Southfield, Michigan 48037
(313) 355-9300

Florist Review (weekly)
Florists' Review Enterprises, Inc.
P.O. Box 4368
Topeka, Kansas 66604
(913) 266-0888

Flowers & (monthly)
Teleflora Plaza, Suite 260
12233 W. Olympic Blvd.
Los Angeles, California 90064
(213) 826-5253

FTD Future Florist Newsletter (3 times/yr)
FTD Headquarters
Education Division
29200 Northwestern Hwy.
Southfield, Michigan 48037
(313) 355-9300

Flower News (weekly)
549 W. Randolph St.
Chicago, Illinois 60606
(312) 236-8648

Marketletter (weekly)
Florist Publishing Company
111 North Canal St., Suite 545
Chicago, Illinois 60606
(312) 782-5505

The Professional Floral Designer (Bimonthly)
American Floral Services, Inc.
P.O. Box 12309
Oklahoma City, Oklahoma 73157

SAF: Business News for the Floral Industry
(monthly)
Society of American Florists
1601 Duke St.
Alexandria, Virginia 22314

State Florist Association Newsletter

Contact your state association for
information

FIGURE 26-5
A partial list of trade publications.

FIGURE 26-6
A state floral design competition sponsored by the FFA.

CERTIFICATION PROGRAMS

Many state florist associations offer state certification programs. These programs attempt to establish knowledge and performance standards in the floral industry and give meaningful credentials to those in the florist industry. Contact your state association to find out if there is a certification program in your state.

Student Activities

1. Make a list of the professional organizations discussed in this chapter and list the major objective of each.
2. Make a list of organizations with student membership.
3. Invite a local florist to talk to the class about professional organizations and trade publications.
4. Write to your state florist association for information on certification programs in your state.
5. Write a letter to a student florist association for information about the organization.

Self-Evaluation

A. Short Answer Questions

1. Why is it important for a florist to be a member of professional organizations and certification programs?

2. Explain the importance of subscribing to professional trade publications.

B. Match the trade organization with the descriptive term that best describes it.

Trade Organizations

_____ 1. Society of American Florists

_____ 2. American Floral Marketing Council

_____ 3. American Academy of Floriculture

_____ 4. Redbook Master Consultants

_____ 5. Allied florists' associations

Descriptive Terms

a. promotes everyday sales of floral products.

b. recognizes excellence in community and floral design service.

c. membership open to all segments of industry.

d. acts as a support system for area florists.

e. improves the floral industry through education.

Appendix **A**

Cut Flowers

Yarrow
Achillea filipendulina
Family: Compositae
Use: Mass
Color: Yellow
Available: Summer
Packaged: Bunch (10)
Vase Life: 7 to 10 days

Lily of the Nile
Agapanthus species
Family: Liliaceae
Use: Form
Color: White, blue
Available: Late spring, summer
Packaged: Bunch (10)
Vase Life: 7 to 14 days

Peruvian Lily
Alstroemeria species
Family: Amaryllidaceae
Use: Form, filler
Color: Orange, pink, yellow, purple, white
Available: Year-round
Packaged: Bunch (10)
Vase Life: 10 days

Anthurium
Anthurium species
Family: Araceae
Use: Form
Color: Red, pink, white
Available: Year-round
Packaged: Individual or bunch (10)
Vase Life: 2 to 3 weeks

Spray Asters
Aster species
Family: Compositae
Use: Filler
Color: Yellow, white, lavender
Available: July through December
Packaged: Bunch
Vase Life: Up to 10 days

Aster
Callistephus chinensis
Family: Compositae
Use: Mass
Color: Blue, lavender, pink, white
Available: Summer, autumn
Packaged: Bunch (25)
Vase Life: 5 to 10 days

Cattleya Orchid
Cattleya species
Family: Orchidaceae
Use: Form
Color: Lavender, white, yellow
Available: Year-round
Packaged: Individual
Vase Life: 5 to 10 days

Celosia
Celosia christata
Family: Amaranthaceae
Use: Mass, form
Color: Red, yellow, orange, pink
Available: Summer, autumn
Packaged: Bunch
Vase Life: 5 to 7 days

Waxflower
Chamelaucium uncinatum
Family: Myrtaceae
Use: line, filler
Color: White, pinks, lavender, bicolors
Available: September through May
Packaged: Bunch
Vase Life: 7 to 10 days

Spray Chrysanthemums: Cushion, Button and Daisy Pompons
Chrysanthemum morifolium
Family: Compositae
Use: Mass, filler
Color: White, yellow, lavender, bronze, orange
Available: Year-round
Packaged: Bunch
Vase Life: 1 to 3 weeks

Tubular Chrysanthemum (Fugi and Spider)
Chrysanthemum morifolium
Family: Compositae
Use: Mass, form
Color: White, yellow
Available: Year-round
Packaged: Bunch
Vase Life: 1 to 3 weeks

Standard Mum, Football Mum
Chrysanthemum morifolium
Family: Compositae
Use: Mass
Color: White, yellow, lavender, bronze
Available: Year-round
Packaged: Individual or bunch
Vase Life: 1 to 3 weeks

Marguerite Daisy
Chrysanthemum frutescens
Family:	Compositae
Use:	Mass, filler
Color:	White, yellow
Available:	Mainly spring, summer
Packaged:	Bunch
Vase Life:	Up to 10 days

Larkspur
Consolida ambigua
Family:	Ranunculaceae
Use:	Line
Color:	Blue, lavender, pink, white
Available:	Spring, summer
Packaged:	Bunch (10 stems)
Vase Life:	7 to 10 days

Cymbidium Orchid
Cymbidium species
Family:	Orchidaceae
Use:	Form
Color:	White, green, pink, yellow
Available:	Year-round
Packaged:	Individual or spray
Vase Life:	1 to 3 weeks

Delphinium
Delphinium elatum
Family:	Ranunculaceae
Use:	Line
Color:	White, blue, lavender, pink
Available:	Year-round
Packaged:	Bunch (10 stems)
Vase Life:	1 week

Standard Carnation
Dianthus caryophyllus
Family: Caryophyllaceae
Use: Mass
Color: Red, white, pink, yellow, bi-colors
Available: Year-round
Packaged: Bunch (25)
Vase Life: Up to 3 weeks

Miniature Carnation
Dianthus caryophyllus nana
Family: Caryophyllaceac
Use: Mass, filler
Color: Purple, white, pink, red, yellow, bicolors
Available: Year-round
Packaged: Bunch
Vase Life: Up to 3 weeks

Lisianthus
Eustoma grandiflorum
Family: Gentianaceae
Use: Mass
Color: Purple, pinks, red, bicolors
Available: May through December
Packaged: Bunch
Vase Life: 1 to 2 weeks

Sunflower
Helianthus species
Family: Compositae
Use: Form, mass
Color: Yellow
Available: Summer, autumn
Packaged: Bunch (10)
Vase Life: 7 to 10 days

Tuberose
Polianthes tuberosa
Family: Agavaceae
Use: Line, form
Color: White
Available: February through October
Packaged: Bunch (10)
Vase Life: 1 to 2 weeks

Protea
Protea species
Family: Proteaceae
Use: Form
Color: Orange, pink, red, purple, yellow
Available: Year-round
Packaged: Individual
Vase Life: 2 to 3 weeks

Freesia
Freesia x hybrida
Family: Iridaceae
Use: Form, mass
Color: Yellow, white, lavender, rose
Available: Year-round
Packaged: Bunch (10)
Vase Life: 7 to 10 days

Gardenia
Gardenia grandiflora
Family: Rubiaceae
Use: Corsage and wedding flower
Color: Creamy white
Available: Year-round
Packaged: Individual or box of 3
Vase Life: 2 to 3 days

Gerbera or Transvaal Daisy
Gerbera jamesonii
Family: Compositae
Use: Mass
Color: Yellow, orange, cream, white, pink, red, purple
Available: Year-round
Packaged: Bunch or individual
Vase Life: 7 days

Gladiolus, Glad
Gladiolus hybrids
Family: Iridaceae
Use: Line
Color: Red, yellow, white, pink, salmon, orange, lavender
Available: Year-round
Packaged: Bunch (10 stems)
Vase Life: 1 to 2 weeks

Baby's Breath
Gypsophila species
Family: Caryophyllaceae
Use: Filler
Color: White
Available: Year-round
Packaged: Bunch
Vase Life: 7 to 10 days

Dutch Iris
Iris species
Family: Iridaceae
Use: Form, mass
Color: Purple, yellow, white
Available: Year-round
Packaged: Bunch (10)
Vase Life: 2 to 6 days

Liatris, Blazing Star
Liatris spicata

Family:	Compositae
Use:	Line
Color:	Lavender, white
Available:	Year-round
Packaged:	Bunch (10 stems)
Vase Life:	7 to 10 days

Rubrum Lily
Lilium rubrum

Family:	Liliaceae
Use:	Form
Color:	Pink
Available:	Spring, summer
Packaged:	Bunch
Vase Life:	10 to 14 days

Asiatic Lily
Lilium species

Family:	Liliaceae
Use:	Form
Color:	Yellow, orange, pink, white
Available:	Year-round
Packaged:	Bunch
Vase Life:	10 to 14 days

Statice
Limonium sinuatum

Family:	Plumbaginaceae
Use:	Filler
Color:	Blue, white, yellow, orange, pink, lavender
Available:	Year-round
Packaged:	Bunch
Vase Life:	Up to 2 weeks

Stock
Matthiola incana
Family: Cruciferae
Use: Line
Color: Purple, white, lavender, yellow
Available: January through October
Packaged: Bunch (10 stems)
Vase Life: Up to 7 days

Bells of Ireland
Molucella laevis
Family: Labiatae
Use: Line
Color: Green
Available: Winter, spring, summer
Packaged: Bunch (10)
Vase Life: 7 to 10 days

Daffodil
Narcissus species
Family: Amaryllidaceae
Use: Form, mass
Color: Yellow, white, bicolors
Available: Winter, spring
Packaged: Bunch (10)
Vase Life: 4 to 6 days

Hybrid Tea Rose
Rosa hybrida
Family: Rosaceae
Use: Mass, form
Color: Red, pink, white, lavender, gold, yellow, bicolors
Available: Year-round
Packaged: Bunch (25)
Vase Life: 3 to 14 days

Sweetheart Rose
Rosa hybrida
Family: Rosaceae
Use: Mass
Color: Red, pink, white, yellow, lavender, bicolor
Available: Year-round
Packaged: Bunch (25)
Vase Life: 3 to 14 days

Snapdragon
Antirrhinum majus
Family: Scrophulariaceae
Use: Line
Color: Yellow, white, pink, lavender, red, orange
Available: Year-round
Packaged: Bunch (10)
Vase Life: Up to 2 weeks

Stephanotis
Stephanotis floribunda
Family: Asclepiadaceae
Use: Corsage work
Color: White
Available: Year-round
Packaged: Package (25)
Vase Life: 3 to 4 days

Calla Lily
Zantedeschia aethiopica
Family: Araceae
Use: Form
Color: White
Available: Year-round
Packaged: Individual or bunch (10)
Vase Life: Up to 10 days

Appendix **B**

Cut Foliages

Ming Fern
Asparagus densiflorus 'Myriocladus'
Family: Liliaceae
Available: Year-round
Packaged: Bunch
Vase Life: Over 2 weeks

Plumosa
Asparagus plumosus
Family: Liliaceae
Available: Year-round
Packaged: Bunch (20–30 stems)
Vase Life: Over 2 weeks

Tree Fern
Asparagus pyramidalis
Family: Liliaceae
Available: Year-round
Packaged: Bunch
Vase Life: Up to 20 days

Sprengeri
Asparagus sprengeri
Family: Liliaceae
Available: Year-round
Packaged: Bunch
Vase Life: 2 weeks

Emerald, Jade
Chamaedorea species
Family: Palmacae
Available: Year-round
Packaged: Bunch (25)
Vase Life: 5 to 7 days

Ti Leaf
Cordyline terminalis
Family: Agavaceae
Available: Year-round
Packaged: Bunch (10)
Vase Life: 1 to 2 weeks

Umbrella Palm
Cyperus alternifolius
Family: Cyperaceae
Available: Year-round
Packaged: Bunch
Vase Life: 3 to 4 weeks

Scotch Broom
Cytisus scoparius
Family: Leguminosae
Available: August through April
Packaged: Bunch
Vase Life: 3 weeks

Eucalyptus
Eucalyptus species
Family: Myrtaceae
Available: Year-round
Packaged: Bunch (by weight)
Vase Life: 10 days

Galax
Galax urceolata
Family: Diapensiaceae
Available: Year-round
Packaged: Bunch (25)
Vase Life: 1 to 2 weeks

Ivy
Hedera helix
Family: Araliaceae
Available: Year-round
Packaged: Bunch
Vase Life: 5 days

Clubmoss
Lycopodium taxifolium
Family: Lycopodiaceae
Available: Year-round
Packaged: Bunch
Vase Life: Over 1 week

Myrtle
Myrtus communis
Family: Myrtaceae
Available: Year-round
Packaged: Bunch
Vase Life: 10 days

Boston Fern, Sword Fern
Nephrolepis exaltata
Family: Polypodiaceae
Available: Year-round
Packaged: Bunch (25)
Vase Life: 5 to 10 days

Pittosporum
Pittosporum tobira
Family: Pittosporaceae
Available: Year-round
Packaged: Bunch
Vase Life: 1 to 2 weeks

Podocarpus, Southern Yew
Podocarpus macrophyllus
Family: Podocarpaceae
Available: Year-round
Packaged: Bunch
Vase Life: 10 days

Leatherleaf, Baker Fern
Rumohra adiantiformis
Family: Polypodiaceae
Available: Year-round
Packaged: Bunch (25)
Vase Life: Over 2 weeks

Italian Ruscus
Ruscus hypoglossum
Family: Liliaceae
Available: Year-round
Packaged: Bunch
Vase Life: 2 weeks

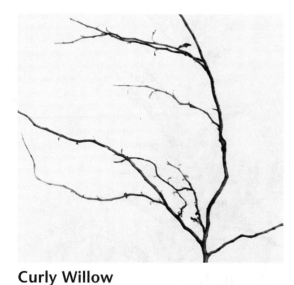

Curly Willow
Salix sachalinensis
Family: Salicaceae
Available: Year-round
Packaged: Bunch
Vase Life: Long lasting, 3 weeks or longer

Huckleberry
Vaccinium ovatum
Family: Ericaceae
Available: Year-round
Packaged: Bunch
Vase Life: 1 to 2 weeks

Bear Grass
Xerophyllum tenax
Family: Liliaceae
Available: Year-round
Packaged: Bunch
Vase Life: 1 to 3 weeks

Appendix **C**

Dried Materials

Button Pompons

Celosia

Eucalyptus
(Glycerin treated)

German Statice

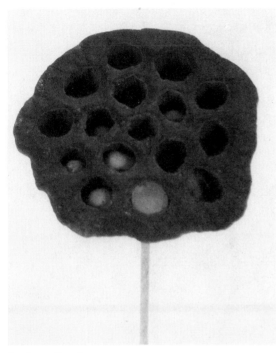

Lotus Pods

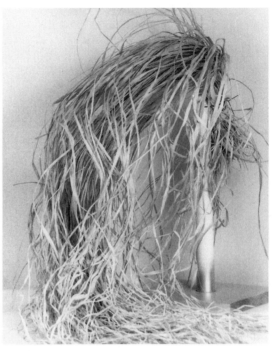

Raffia

Reindeer Moss

Sea Grape

Spanish Moss

Poppy Seed Heads

Wheat

Yarrow

Appendix **D**

Foliage Houseplants

Chinese Evergreen (Silver Queen)
Aglaonema crispum
Light: Low
Water: Moist/dry

Aloe
Aloe vera
Light: High
Water: Dry in winter

Norfolk Island Pine
Araucaria excelsa
Light: Medium
Water: Moist/dry

Coral Berry
Ardisia crispa
Light: Medium
Water: Moist but not wet

Asparagus Fern
Asparagus plumosus
Light: Medium
Water: Moist/dry

Bird's Nest Fern
Aspienium nidus
Light: Medium
Water: Moist but not wet

Ponytail
Beaucarnea recurvata
Light: Medium
Water: Moist/dry

Rex Begonia
Begonia rex
Light: Medium
Water: Moist/dry

Caladium
Caladium species
Light: Medium
Water: Moist but not wet

Fishtail Palm
Caryota mitis
Light: Medium
Water: Moist/dry

Grape Ivy
Cissus rhombifolia
Light: Medium
Water: Moist but not wet

Croton
Codiaeum variegatum pictum
Light: High
Water: Moist but not wet

Jade Plant
Crassula argentea
Light: High
Water: Moist/dry

Dumbcane
Dieffenbachia maculata 'camille'
Light: Medium
Water: Moist/dry

False Aralia
Dizygotheca elegantissima
Light: Medium
Water: Moist but not wet

Corn Plant
Dracaena fragrans
Light: Medium
Water: Moist but not wet

Madagascar Dragon Tree
Dracaena marginata
Light: Medium
Water: Moist but not wet

Pineapple Dracaena
Dracaena deremensis 'compacta'
Light: Medium
Water: Moist but not wet

Warneckei Dracaena
Dracaena dermensis 'warneckei'
Light: Medium
Water: Moist but not wet

Gold Dust Dracaena
Dracaena godseffiana
Light: Medium
Water: Moist but not wet

Weeping Fig
Ficus benjamina
Light: Medium
Water: Moist/dry

Rubber Plant
Ficus elastica decora
Light: Medium
Water: Moist/dry

Fiddle Leaf Fig
Ficus lyrata
Light: Medium
Water: Moist/dry

Purple-Passion Vine, Velvet Plant
Gynura species
Light: Medium
Water: Moist but not wet

English Ivy
Hedera helix
Light: Medium
Water: Moist but not wet

Polka-Dot Plant
Hypoestes phyllostachya
Light: Medium
Water: Moist but not wet

Prayer Plant
Maranta species
Light: Medium
Water: Moist/dry

Split-Leaf Philodendron
Monstera deliciosa
Light: Medium
Water: Moist/dry

Parlour Palm
Neanthe bella
Light: Low
Water: Moist but not wet

Blushing Bromeliad
Neoregelia carolinae
Light: Medium
Water: Moist but not wet

Boston Fern
Nephrolepis exaltata
Light: Medium
Water: Moist/dry

Emerald Ripple
Peperomia caperata
Light: Medium
Water: Moist/dry

Variegated Peperomia
Peperomia obtusifolia varigata
Light: Medium
Water: Moist/dry

Watermelon Peperomia
Peperomia sandersii
Light: Medium
Water: Moist but not wet

Lacy Tree Philodendron
Philodendron selloum
Light: Medium
Water: Moist but not wet

Heart Leaf Philodendron
Philodendron scandens
Light: Low
Water: Moist but not wet

Song of India
Pleomele reflexa
Light: Medium
Water: Moist/dry

Moses-in-the-Cradle
Rhoeo spathacea
Light: Medium
Water: Moist but not wet

Snake Plant, Mother-in-law's Tongue
Sansevieria trifasciata
Light: Medium
Water: Moist/dry

Umbrella Tree
Schefflera actinophylla
Light: Medium
Water: Moist but not wet

Devil's Ivy, Golden Pothos
Scindapsus aureus
Light: Medium
Water: Moist/dry

Goose Foot Plant, Arrowhead Vine
Syngonium podophyllum
Light: Medium
Water: Moist but not wet

Appendix **E**

Permanent Flowering Potted Plants

Flamingo Flower
Anthurium scherzerianum
Light: Medium
Water: Moist but not wet

Bougainvillea, Paper Flower
Bougainvillea glabra
Light: Bright
Water: Moist/dry

Corsage Orchid
Cattleya hybrid
Light: High
Water: Moist but not wet

Rose of China
Hibiscus rosa sinensis
Light: High
Water: Moist but not wet

Kalanchoe
Kalanchoe blossfeldiana
Light: High
Water: Moist/dry

Window Box Oxalis
Oxalis rubra alba
Light: High
Water: Moist/dry

Lollipop Plant
Pachystachys lutea
Light: Medium
Water: Moist/dry

Geranium
Pelargonium hortorum
Light: High
Water: Moist/dry

Azalea
Rhododendron hybrid
Light: Medium
Water: Moist but not wet

African Violet
Saintpaulia ionantha
Light: Medium
Water: Moist but not wet

Peace Lily
Spathiphyllum wallisii
Light: Medium
Water: Moist but not wet

Christmas Cactus
Zygocactus truncatus
Light: Medium
Water: Moist but not wet
        Moist/dry during fall

Temporary Flowering Potted Plants

Amaryllis
Hippeastrum hybrida
Light: High
Water: Moist/dry

Rieger Begonia
Begonia x hiemalis
Light: Medium
Water: Moist but not wet

Pocketbook Plant
Calceolaria herbeohybrida
Light: Medium
Water: Moist/dry

Chrysanthemum
Chrysanthemum morifolium
Light: High
Water: Moist but not wet

Cyclamen
Cyclamen persicum
Light: Medium
Water: Moist but not wet

Poinsettia
Euphorbia pulcherrima
Light: High
Water: Moist/dry

Gerbera Daisy
Gerberia jamesonii
Light: High
Water: Moist/dry

Dutch Hyacinth
Hyacinthus orientalis
Light: Medium
Water: Moist but not wet

Hydrangea
Hydrangea macrophylla
Light: Medium
Water: Moist but not wet

Easter Lily
Lilium longiflorum
Light: Medium
Water: Moist but not wet

Lisianthus
Lisianthus russellianus
Light: Medium
Water: Moist/dry

Primula
Primula acaulis
Light: Medium
Water: Moist but not wet

Cineraria
Senecio cruentus
Light: Bright but sunless
Water: Moist but not wet

Gloxinia
Sinningia speciosa
Light: Medium
Water: Moist but not wet

Tulip
Tulip species
Light: Medium
Water: Moist but not wet

Flower Arrangement Rating Scale

Directions: Indicate the degree to which this student has utilized the following principles of flower arrangement by placing a check anywhere along the horizontal line under each item.

	5　　　　4	3	2　　　　1	Points
Design	Shows high quality of design. Needs little or no improvement.	Shows some use of design. Needs improvement.	Shows little or no use of design. Needs much improvement.	
Balance	Shows visual and actual balance. Needs little or no improvement.	Shows attempt at balance. Needs some improvement.	Not balanced. Needs much improvement.	
Proportion	Container and flowers are in good propor-tion. Needs no improvement.	Shows attempt at proportion. Needs some improvement.	Shows lack of proportion. Needs much improvement.	
Harmony	Component parts blend together. Needs little or no improvement.	Component parts blend together some-what. Needs some improve-ment.	Component parts do not blend together.	
Focal Point	Definite estab-lishment of focal points. Needs little or no improvement.	Focal point established but needs some improvement.	Focal point not established. Needs much improvement.	
Rhythm	Shows good use of rhythm. Eye moves smoothly through arrange-ment. Needs little or no improvement.	Shows some use of rhythm but needs improve-ment.	Shows little or no use of rhythm. Needs much improve-ment.	

	5	4	3	2	1	Points
Repetition	Shows use of repetition to add interest to the arrangement.		Shows use of repetition but needs some improvement.	Incorrect use of repetition. Needs much improvement.		
Unity	Whole composition is a complete unit. Needs little or no improvement.		Shows use of unity but needs improvement to become a complete unit.	Complete lack of unity. Needs much improvement.		

Directions: Indicate the degree to which this student has appropriately used color in this arrangement by placing a check anywhere along the horizontal line under each item.

	5	4	3	2	1	Points
Color	Combination of color is pleasing and appropriate for occasion.		Combination of color needs improvement.	Poor use of color or is not suitable for occasion.		

Directions: Indicate your overall rating of the salability of the design on the following scale by placing a check anywhere along the horizontal line.

	5	4	3	2	1	Points
Salability	Excellent salability. Better than average of retail flower shops.		Shows good salability. Average for most retail shops.	Poor salability. Not suitable for sale in flower shop.		

Floral Design-Related Web Sites

American Floral Services (AFS)
www.afs.com

Best Buy Floral Supply
www.bestbuyfloral.com

Botanique Preservation Equipment
www.botaniquefrzdry.com

Carik Services, Inc.
www.carikonline.com

FTD Association
www.ftdassociation.org

Florafax International
www.florafax.com

Floralife, Inc.
www.floralife.com

Flowers, Inc. Balloons
www.flowerincballoons.com

Lion Ribbon Co./C.M. Offray
www.offray.com

Pacific Wire & Supply, Inc.
www.pac-wire.com

Rainforest Preserved Plants
www.islandnet.com/~hapac

Smithers-Oasis U.S.A.
www.smithersoasis.com

Schusters of Texas, Inc.
www.schustersoftexas.com

America's Florist
www.americasflorist.com

Balloon Supply of America
www/balloonsupplyofamerica.com

California Cut Flower Commission
www.ccfc.org

Colors From the World
www.cfw.com

Floracubes
www.floracubes.com

Floral Concepts International
www.floralcommunications.com

Florists' Transworld Delivery (FTD)
www.ftd.com

The John Henry Company
www.jhc.com

Mitchell Wreath Rings
www.mitchellwreaths.com

Pioneer Balloon Company
www.qualatex.com

Rosenfeld Ribbon
www.rosenfeldribbon.com

Society of American Florists (SAF)
www.safnow.org

Glossary

Accessory: A nonessential addition to any floral arrangement.

Allied florists' associations: Groups from within a region of a state that act as a support system for a florist in a particular area.

American Academy of Floriculture: A branch of the Society of American Florists whose purpose is to encourage and recognize excellence in community and floral industry service.

American Federal Period: The political, social, and decorative formation era in America following the Revolutionary War. Floral arrangements during this period were styled after ancient classic designs as well as elaborate European massed, symmetrical bouquets.

American Floral Marketing Council: Sponsored by the Society of American Florists but funded separately. AFMC's purpose is to promote everyday sales of floral products.

American Institute of Floral Designers: A trade organization dedicated to establishing higher standards in professional floral design.

Analogous Harmony: A color scheme that utilizes three or more hues next to each other on the color wheel.

Anchor Tape: Waterproof tape used primarily to hold floral foam in place.

Antitranspirants: Materials that slow down the loss of water from the flower through transpiration.

Arm Bouquet: A tied cluster of flowers carried across the forearm used for weddings or as a presentation bouquet.

Artistic Displays: A display designed to focus customer attention. Artistic displays can be used to draw customers to a "dead" area of store space, to stop shoppers at an important section, and to create an image.

Bactericide: An agent added to preservatives that kills bacteria in the water.

Balance: A design principle. The placement of objects to create a physical and visual feeling of stability in a design.

Baroque Period: Artistic period in Europe characterized by elaborate and massive decorative elements and curved rather than straight lines. Floral arrangements in this period were tightly massed and overflowing, displaying a rhythmic asymmetrical balance.

Binomial Name: A naming system (meaning two names) for plants using the genus taxon and the species taxon.

Bolts: Ribbon is purchased on cardboard spools called bolts.

Botrytis: A fungus that causes brown spots on flower petals.

Boutonniere: A single flower or several small flowers worn by a man on his lapel.

Bud Vase: A tall narrow container with a small neck designed to hold one flower or a number of flowers.

Buying Up: A customer buys a more expensive item than planned.

Byzantine Period: A.D. 320–600. Artistic period in southeast Europe and southwest Asia in which floral designs were distinguished by height and symmetry. Symmetrical stylized tree compositions were introduced during this period.

Calyx: The outer whorl of protective leaves, or sepals (flower bud leaves), of a flower.

Cardett: A plastic stem with a three-prong holder at the top to hold a card.

Carriage Trade Shop: A flower shop that caters to an elite clientele such as wealthy private party and corporate accounts.

Cascading Bouquet: A handheld wedding bouquet style in which the flowers hang down (cascade) below the main portion of the design.

Casket Cover: A floral arrangement that is placed over the closed lid of a casket.

Chaplet: A wreath or garland for the head, usually made from flowers and foliage. Introduced during ancient periods.

Clearinghouse: Essentially a bookkeeping service which ensures the sending and receiving florist that each shop will receive proper payment for orders transmitted and delivered. A wire service is a clearinghouse for floral orders.

Colonial Bouquet: A handheld bouquet constructed in a circular shape with an attached handle.

Colonial Williamsburg Period: A.D. 1714–1820. Artistic period in America characterized by bouquets that were rounded and massed, often combining fresh and dried flowers.

Common Name: The name by which a plant is known by the people living in an area rather than by its scientific name.

Complementary Harmony: A combination of any two colors opposite each other on the color wheel. This combination produces a strong contrast.

Conditioning Flowers: The technique of treating flowers to extend their life.

Container: Anything that holds water, but also helps express the idea the designer has in mind for an arrangement.

Control Board: A visual control over the day's deliveries. Each delivery ticket is put on the proper day's hook so that the delivery person knows in advance what orders must go where on what day.

Cornucopia: A basket shaped like a horn or cone overflowing with fruit and vegetables, flowers and grain. Introduced during the Greek period, the cornucopia is the symbol for abundance and is often used for Thanksgiving decorations.

Corsage: Flowers worn by women.

Curling Ribbon: A special ribbon made from polypropylene that curls when scraped with a knife or scissors and can be used to tie latex or Mylar balloons.

Delivery Pool: A group of shops pooling delivery resources and personnel. Each shop brings its delivery items to a central meeting point at a specified time. The deliveries are divided by areas, and each shop then takes all of the deliveries for its area.

Desiccant-Drying: Burying flowers in a substance that will extract moisture from the flowers by absorption.

Design: A planned organization of plant, floral, or accessory materials for a specific purpose.

Designer: A person who has the ability to arrange flowers and plant material in an artistic manner and has an understanding of the principles of design.

Designer's Assistant: A person who is training to become a designer and usually works with a designer.

Display: The visual and artistic aspects of presenting a product to the customer.

Display Elements: The components used in the process of assembling displays.

Divisional Percentage Pricing: Includes net profit as a factor in pricing along with operating expenses, labor, and cost of goods.

Dixon Picks: Two wooden picks attached on opposite ends of a flexible metal strip; used as a mechanical aid in floral work, particularly large sympathy designs.

Double-Faced Ribbon: Ribbon that has the same finish on both sides.

Drenching: Watering plants until water runs out the bottom of the pot.

Dripless Candles: Dripless candles are made with a metal casing in the shape of a candle. Inside the casing, a spring pushes a thin wax candle up to the top as it burns. This eliminates the problem of dripping wax.

Dutch-Flemish Period: A.D. 1600–1750. An artistic period in Europe in which flower arrangements were copied from the paintings of Dutch and Flemish artists of the period. Typical floral arrangements are massed and overflowing with the use of many varieties and colors of flowers.

Early American Period: A.D. 1620–1720. Early period in America in which floral arrangements were simple, using native plant materials such as wildflowers, weeds, and grains. These were usually placed in simple utility items in the home.

Egyptian Period: 2800–32 B.C. The floral style of this period was simplistic, repetitious and highly stylized. Flowers and fruits were placed in carefully alternating patterns. Chaplets, wreaths, and garlands were also popular.

Emphasis: The creation of visual importance or accent in a design.

Employability Portfolio: A collection of documents that shows you have knowledge, mastery, and job readiness in a particular occupational area.

Empire Period: A.D. 1804–1814. Artistic period in France during the rule of Napoleon Bonaparte. Floral arrangements were massive in size and weight and very masculine. Bouquets were more compact than earlier French periods with simple lines in a triangular shape with strong emphasis on color.

English-Georgian Period: A.D. 1714–1760. Also called **Georgian**, refers to the period in England during the reigns of George I, II, and III. Floral styles included handheld bouquets. Arrangements were symmetrical and contained a great variety of fragrant flowers.

Ethylene Gas: A naturally occurring gas in flowers that hastens maturity and causes rapid deterioration of cut flowers.

Faience: Glazed earthenware containers made of finely ground silicate.

Filler Flowers: Filler flowers are used to fill in the gaps between mass flowers and to give depth to a design.

Floral Adhesive: A rubber cement that has been developed for use on fresh flowers. The harmful chemicals have been removed from floral adhesive and it may be used in designing corsages where very light and delicate flowers are used.

Floral Clay: A waterproof, sticky material used to fasten anchor pins that hold floral foam in place or to anchor pinpoint holders firmly in place.

Floral Foam: A soft, light-weight material capable of absorbing large quantities of water. The most commonly used material for holding the stems of flowers in an arrangement.

Floral Preservative: Chemicals added to the water in a vase to aid in extending the life of cut flowers in an arrangement.

Florist Foil: A plastic-coated aluminum foil used to cover pots.

Florist Shears: A tool used for cutting flower stems, wires, and ribbons in a flower shop.

Focal Point: The area of a design that attracts and holds the interest of the viewer.

Form: The shape or silhouette of an arrangement or the outline that the arrangement projects against a space.

Form Flowers: Form flowers have unusual, distinctive shapes and add emphasis to an arrangement.

Franchise Shop: A shop purchased from a parent company and operated according to the company's rules and regulations.

Freeze-Dried Flowers: Flowers that have had all of the moisture mechanically removed from their cell structures.

French Rococo Period: During this period in France, floral designs were formal and feminine. Arrangements were asymmetrical and curvilinear in form. Designs were delicate and airy.

Full-Service Shop: A traditional type of retail flower shop offering a wide variety of services and products.

Gauge: Florist wire comes in various weights and diameters called gauge. Wire gauges range from number 18 (thickest) to 32 (thinnest).

Glycerin: A colorless liquid made from fats and oils that can be used to preserve foliages and some flowers.

Greek Period: 600–46 B.C. Floral designs of this time were garlands and wreaths. Flowers were scattered on the ground during festivals. Fragrance and symbolism were important.

Greening Pins: Wire pins with a U-shaped top that are used as fastening devices.

Greeting Approach: A sales approach in which the salesperson simply welcomes the customer to the store using the customer's name if possible.

Harmony: Harmony in a design refers to a blending of all components of the design to create a pleasing relationship of color, texture, shape, size, and line so that a central idea or theme is accomplished.

Heat Sealer: An electric device used to seal the stems of Mylar balloons.

Helium: A nontoxic, nonflammable gas that is lighter than air.

Hi Float: A nontoxic sealant used to treat the interior of latex balloons.

Holiday Deliveries: Deliveries during holiday or peak times when extra personnel are hired.

Hot Glue: An adhesive material used extensively in the florist shop.

Hydration: Hydration is the process of plant's capillaries carrying water and nutrients up the stem to the leaves and flowers. Hydration is what keeps cut flowers fresh.

Ikebana: The art of Japanese floral arrangement; literally meaning "giving life to flowers."

Impulse Buying: A customer sees an item and buys it even though the item was not what the customer originally intended to buy.

Intensity: The brightness or dullness of flower color.

Jardiniere: A decorative container used to hold flowers that can be made from a number of materials such as pottery, china, plastic, and brass.

Last-Minute Deliveries: Crisis or emergency special service deliveries in which the delivery must be made immediately.

Latex: A material made of rubber and other chemicals used to make balloons that will stretch when filled with compressed air or helium.

Leader Pricing: A strategy used to help lure customers into the flower shop by pricing items well below their normal price.

Line: The visual path the eye follows to create motion in the design and the framework holding the entire arrangement together.

Line Arrangement: The Japanese designs are characterized by minimum use of plant material and the careful placement of branches and flowers. Each placement has meaning as does the angle of placement.

Line Design: A general term for a floral arrangement characterized by strong lines with negative space between for emphasis.

Line Flowers: Long, slender spikes or blossoms with florets blooming along the stem. Bare twigs or other similar type materials are also classified as line flowers. These flowers are used to establish the skeleton or outline of an arrangement.

Line-Mass Arrangement: The contemporary design style that combines the linear shapes with the massing of flowers at the focal point. Most floral design in the United States is referred to as line-mass and combines Oriental and European ideas.

Mass Arrangement: The European floral design style in which large, closely spaced flowers are located at the edge of the design. Large quantities of flower types and colors are used to create a massive display.

Mass Flowers: Mass flowers are single-stem flowers with large rounded heads and are used inside the framework of the linear flowers toward the focal point.

Mass-Market Shop: A shop located in a general merchandise chain store or grocery store. These shops are cash-and-carry operations and do not offer delivery service.

Mechanics: All of the materials used to assist the designer in placing and holding flowers.

Merchandise Approach: In the merchandise approach the salesperson makes a comment about an item that has the customer's attention.

Middle Ages: A.D. 476–1600. The period of European history between ancient and modern times. Also known as the Dark Ages. Little is known of floral art during this period.

Monochromatic Harmony: A color scheme consisting of a single hue along with variations of this color in tints, tone, and shades.

Mylar Balloons: Balloons that are produced from a thin metallic film that will not stretch.

Negative Space: Empty areas between flowers or materials.

Nesting Pricing: Nesting pricing is when a collection of goods are sold at a single price.

Net Tufts: A corsage accessory made from tulle with an extended wire stem.

Nosegay: A tight grouping of flowers and foliages in a handheld bouquet. Also called a tuzzy-muzzy or posie bouquet.

Occidental Style: A loose term referring to the mass designs that originated in Europe that were characterized by the use of large numbers of flowers.

Oriental Style: A loose term referring to line designs. The oriental style uses few materials and emphasizes simplicity, form, line, and texture.

Out-of-Area Deliveries: Deliveries made beyond the shop's normal delivery area.

Paper Flowers: Paper flowers are created from rice paper, parchment, and bark fiber paper.

Petiole: The leaf stem.

pH: The pH is a measure of how acidic or basic a water is on a scale of 0 to 14. Seven on the scale is neutral.

Photosynthesis: A process of production of food in the leaves of a plant by the action of light.

Picking Machine: A device used to attach metal picks to flower stems.

Picks: Picks can be wooden or steel and are attached to flower stems. They are very helpful in making wreaths as well as dried and artificial arrangements.

Plant Face: The most attractive side of a plant.

Polychromatic Harmony: Polychromatic harmony uses three or more unrelated colors.

Pot-et-fleur: The resulting product of adding cut flowers to a dish garden for temporary addition of color.

Product-Oriented Displays: Product-oriented displays focus on the direct promoting of merchandise.

Professional Floral Commentators–International: An organization whose purpose is to help educate all segments of the industry on innovations in floral design, floral marketing, merchandising, and fresh flower care and handling. They also introduce new products and floral varieties and seek to improve commentating at all floral events.

Progression: A gradual change in an arrangement by increasing or decreasing one or more qualities.

Proportion: The relationship of all parts of an arrangement to each other.

Pruning Shears: Shears useful in cutting heavy stems that are too large to be easily cut by a knife or florist shears.

Radiation: An attempt to make all stems appear to come from one central axis.

Ratio Markup Plus Labor: Ratio markup pricing is used with different ratios for different products according to whether nonperishable, perishable or items requiring a lot of special attention are involved.

Redbook Master Consultants: A nonprofit corporation organized to improve the floral industry through education. Its main objective is to provide needed educational programs to florists in the area where they live and work.

Regular Deliveries: Deliveries handled through the regular daily deliveries within the shop's established delivery area.

Regulator: A valve attached to a helium tank to dispense helium into balloons.

Renaissance Period: A.D. 1400–1600. A period in Europe following the Middle Ages that saw a rebirth of many interests, particularly in the arts. Floral arrangements during this period were massed in tight symmetrical shapes.

Repetition: A method of obtaining rhythm by repeating similar elements throughout a design.

Respiration: An intracellular process in which food is oxidized with the release of energy. The complete breakdown of sugar or other organic compounds to carbon dioxide and water.

Resume: A list of your experiences relating to the kind of job you are seeking.

Rhythm: The movement of the eye through a design toward or away from the center of interest. It is the flow of line, textures, and color that expresses a feeling of motion.

Ribbon Shears: Shears (scissors) useful in cutting ribbons and decorative foils used by the florist.

Roman Period: 28 B.C.–A.D. 325 During this period, flowers were used to make garlands and wreaths. The use of plant material was more elaborate than in previous periods.

Salesperson: A person who possesses skills in the art of selling.

Satin Ribbon: Ribbon made of a glossy fabric; probably the most frequently used ribbon in the flower shop.

Scale: The relationship between an arrangement and the area where the arrangement is to be displayed.

Scientific Name: The scientific name of a plant is based on a classification system according to how plants are related to each other. This system was developed by a Swedish botanist named Carolus Linnaeus in 1743.

Selection Guide: A selection guide has color pictures of a wide variety of arrangements and is used to communicate design styles and customer preferences to the florist.

Service Approach: In the service approach, the salesperson asks the customer if he or she needs assistance; best used when it is obvious the customer is in a hurry.

Shade: The result of adding black to a color.

Silica Gel: An industrial compound that may be purchased for drying flowers. It is probably the best drying agent for preserving flowers because it dries quickly and the flowers retain more of their natural colors.

Single-Faced Ribbon: Ribbon having a shiny finish on one side and a dull finish on the other side.

Society of American Florists: The only national trade association representing the needs and interest of all segments of the floral industry.

Specialty Shop: A shop that targets a particular need in the market by specializing in one segment of the industry.

Speed Covers: Preformed aluminum foil pot covers available in a variety of colors and sizes.

Split Complementary Harmony: Split complementary harmony uses any color with the two colors on each side of the complement.

Standard Ratio Markup Pricing: A flexible system of pricing that operates by determining the wholesale cost of an item and then multiplying that cost by a number from 2 to 4 or more to cover operating cost, labor, and profit.

State Florist Associations: State florist associations are devoted to the support and improvement of floral-related businesses within the state. They sponsor educational seminars and workshops on a variety of topics and often sponsor design competition among the florists throughout the state.

Steel Pick: Metal picks sold in lengths from 1 3/4 inches to 3 inches. They are attached to the stems of various materials with a steel pick machine.

Stem Shop: Shops that are cash-and-carry operations and offer a wide variety of flowers by the single stem or bunch.

Stem Wrap or Floral Tape: A nonsticky tape that sticks to itself when stretched. It is used to cover wires, bind wires to flower stems, and to bind wired and taped flowers together.

Stomata: The tiny openings on the bottom of the leaves of a plant.

Studio Operation: A shop that concentrates on specialty and party work for an exclusive client base. It may operate out of a warehouse without a storefront since most sales are made on location.

Succulent: The portion of the stem which is soft and juicy and therefore a better conductor of water.

Suggestion Selling: Suggestion selling encourages the customer to purchase additional merchandise related to the original purchase.

Taping: The process of binding wires or flowers and accessories together using florist tape.

Taxa: A group of plant categories.

Taxon: A category of plants.

Texture: The physical surface appearance that an object projects.

Theme Displays: Displays based on a specific subject or topic such as a Christmas display.

Tint: The result of adding white to a color.

Tone: The result of adding gray to a color.

Topiary: An evergreen tree or shrub that has been trimmed or trained to an unnatural shape. A floral design to create this appearance, such as a topiary ball or cone.

Total Dissolved Solids: A measurement of the dissolved salt level in water.

Trade Organization: A nonprofit organization whose purpose is to assist its members through educational programs and to promote the industry.

Transition: Making a gradual change in a flower arrangement by the blending of colors, line patterns, and textures.

Transpiration: The process by which gases, including water vapor, move from an area of greater concentration to an area of lesser concentration.

Triadic Harmony: A color harmony using any three colors equally spaced on the color wheel.

Tulle: A type of decorative netting used as an accessory in corsage and wedding work.

Unity: Unity in a design is achieved when all the parts of the design suggest a oneness in idea or impression by repeating the same flower and colors throughout.

Upselling: Convincing a customer to purchase an item of higher value.

Value: How light or dark a color is.

Victorian Period: A.D. 1820–1901. Period named for Queen Victoria. Floral arrangements are characterized as being massive, overdone, and flamboyant.

Visual Merchandising: The manner in which a florist achieves the goal of selling products or services.

Water Picks: A plastic or glass tube which is topped with a rubber stopper and holds water. Flower stems are inserted through the stopper so that the stem may reach the water in the tube.

Wholesale Florist: The wholesaler purchases goods from around the world and sells to the retail florist rather than to the general public.

Wire Cutters: A tool used for cutting wires and the stems of artificial flowers which contain a wire.

Work Sample: Copies of student work that include steps taken to complete the floral project, workplace skills demonstrated by this project, and ways to improve the work sample.

Xylem: Water-conducting tubes in the stem of plants.

Zoning System: A zoning system groups orders together according to delivery runs.

Index

Teleflora, 399(fig.)
Tell-Tale drying agent, 330–331
Temperature
  and flower preservation, 83
  for potted plants, 377
Temporary flowering potted plants, 357,
    357(fig.)
  illustrated, 481–485
Temporary indoor potted plants, 357,
    357(fig.)
Tertiary color, 55
Texture, 52, 54, 54(fig.), 115
Thanksgiving arrangement, 244–248
  table arrangement, 245–248,
    245–247(figs.)
Theme displays, 411–412, 411(fig.)
  monthly planning calendar, 412(fig.)
Ti leaf, 455(ill.)
Tint, 55
Tone, 55
Topiaries, 251–256
  constructing, using fresh evergreens,
    254–257, 254–257(figs.)
  constructing, using preserved foliages,
    251–254, 251–253(figs.)
Total dissolved solids (TDS), 75
Trade organizations, 432–437, 433(fig.)
  AAF (American Academy of Floriculture),
    433
  AFMC (American Floral Marketing
    Council), 434
  AIFD (American Institute of Floral
    Designers), 434
  Allied florists' associations, 435
  PFCI (Professional Floral Commentators—
    International), 434
  RMC (Redbook Master Consultants),
    434–435, 435(fig.)
  SAF (Society of American Florists), 432
  state florist associations, 435–436
  wholesalers' open houses and design
    schools, 436–437, 436(fig.)
Trade publications, 437
  list, 438(fig.)
Training for a job, 10–11

Transition, 49, 51
Transpiration, 72
Transvaal daisy, 448(ill.)
Tree fern, 454(ill.)
Triadic harmony, 58, 58(fig.)
Triangular arrangements
  asymmetrical triangle, 185–188, 186(fig.)
  centerpiece designs, 183–185, 184(fig.)
  equilateral triangle, 178–181, 178(fig.)
  isosceles triangle, 181–183, 181(fig.)
  right triangle, 190–193, 190(fig.)
  scalene triangle, 188–189, 189(fig.)
Tuberose, 447(ill.)
Tubular chrysanthemum (fugi and spider),
    444(ill.)
Tulip, 485(ill.)
  in mound design, 156
  wire size, 104
Tulle, 135
Tussie-mussies, 35

Umbrella palm, 455(ill.)
Umbrella tree, 475(ill.)
Underwater stem cutter, 77–78, 79(fig.)
Unit cost of goods, determining, 382–383
Unity, 51
Upselling, 394

Valentine mug arrangement, 237–238,
    238(fig.)
Valentine's Day arrangements, 235–238
  Carnation in a Coke can, 236–237,
    236–237(figs.)
  Valentine mug arrangement, 237–238,
    238(figs.)
Value (color), 55
Variegated peperomia, 474(ill.)
Variety display, 413, 414(fig.)
Velvet plant, 471(ill.)
Vertical arrangement, 201–203, 201(fig.)
  constructing an arrangement, 201–203,
    202–203(figs.)
Victorian period, 35, 154
Visual merchandising, 406, 407(fig.)